Marina T. Stojanova
Dragutin A. Djukic
Monika Stojanova
Aziz Şatana

Nutrition, Quality and Biological Potential of Grapevine

www.novapublishers.com

DOI: https://doi.org/10.52305/ZUEH8540

Library of Congress Cataloging-in-Publication Data

ISBN: 979-8-89113-843-8 (softcover)
ISBN: 979-8-89113-918-3 (e-book)

Published by Nova Science Publishers, Inc. † New York

The vineyard needs a servant, while the wine requires a master. The first sip of wine is akin to a first kiss – it leaves you yearning for another. So, let yourself indulge in the pleasures of drinking wine and relishing in the joy and laughter it brings.

André Tchelistcheff

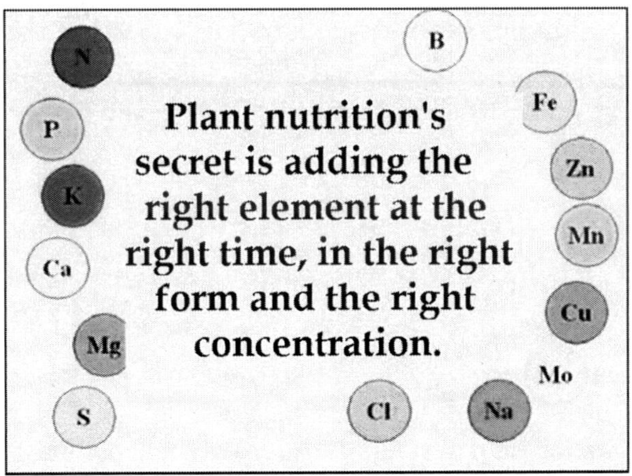

Contents

Preface ix

Chapter 1 **Physiological Role, Symptoms of Deficiency and Excess of Macro and Microelements in Grapevine Nutrition** 1

1. Basic Explanation.. *1*

2. Historical Development of the Grapevine *2*

3. Significance of the Vineyard Growing.. *4*

3.1. Significance of Fertilization in Viticulture Production.. *7*

4. Utilization, Role and Significance of Nutrient Elements in Grapevine Nutrition.. *14*

4.1. How Does the Grapevine Take Up Nutrients? *16*

4.2. Movement of Nutrients in the Grapevine.. *20*

4.3. Factors Affecting the Accumulation of Ions.. *21*

4.4. Balance and Antagonism of Individual Nutritional Elements.. *26*

5. Physiological Role of Macroelements *28*

5.1. Carbon (C), Oxygen (O) and Hydrogen (H) *29*

5.2. Nitrogen (N) *29*

5.3. Phosphorus (P_2O_5) *38*

5.4. Potassium (K_2O).. *42*

5.5. Calcium (Ca).. *47*

5.6. Magnesium (Mg) *50*

5.7. Sulphur (S).. *54*

Appendix A: Symptoms of Deficiency and Excess of Macroelements (Stojanova, 2023).. *56*

6. Physiological Role of Microelements *61*
6.1. Iron (Fe) *65*
6.2. Manganese (Mn) *68*
6.3. Copper (Cu) *71*
6.4. Boron (B) *74*
6.5. Zinc (Zn) *78*
6.6. Cobalt (Co) *82*
6.7. Molybdenum (Mo) *82*
Appendix B: Symptoms of Deficiency and Excess of Microelements (Stojanova, 2023) *85*

Chapter 2 Influence of Macro and Microelements on the Quality and Biological Potential of Grapes and Wine 91
1. Basic Explanation *91*
2. Chemical Composition of Grapes *92*
2.1. Carbohydrates in Grapes *96*
2.2. Acids in Grapes *98*
2.3. Enzymes *101*
2.4. Mineral Matters *102*
2.5. Nitrogen Compounds *104*
2.6. Lipids *105*
2.7. Carotenoids *106*
2.8. Pectin Substances *107*
2.9. Volatile Components of Grapes That Affect the Aroma *107*
2.10. Vitamins *107*
2.11. Waxes *108*
3. Polyphenolic Compounds in Grapes *108*
3.1. Phenolic Compounds *109*
3.2. Flavonoids *116*
3.3. Flavonols *117*
3.4. Flavan-3-ols *118*
3.5. Proanthocyanidins *119*
3.6. Anthocyanins *119*
3.7. Tannins *124*
4. Coloured Substances in Grapes *129*
4.1. Coloured Substances of Black Grapes and Red Wines *130*
4.2. Coloured Substances on White Grapes *135*

5. Aromatic Compounds *135*
6. Phenolic Compounds in Wine and the Biological Potential of Wine *138*

References 149

About the Authors 169

Index 173

Preface

Grapes are considered one of the most important food products in the world, both because of the high content of phytonutrients that have a beneficial effect on the health of the human population, and because they can be used to obtain numerous products and by-products, which have their application in various industries.

Used fresh as fruit or in the form of juice, wine, raisins and jam, grapes are among the highest-quality fruits for consumption around the world. Grapes are rich in vitamins (A, B, C, E, and P), minerals, carbohydrates, lipids, proteins and polyphenolic compounds distributed among the skin, seeds and pulp. Among other ingredients, grapes contain nitrogenous substances, aromatic substances, coloured substances, tannins, polyphenols, etc. Medicinal properties of substances from the group of polyphenols (resveratrols), which are found in the seeds and skin, mostly in the black varieties, have been proven. Polyphenols are antioxidants that eliminate harmful substances from the human body, help the body in the use of vitamin C, act preventively, slow down the ageing of cells, and have antibacterial and antiviral effects.

By applying modern agrotechnical measures, with a special focus on nutrition, that is, by applying optimal doses of macro and microelements in an appropriate period, it is possible to significantly influence the development of the grapevine, and thus the production of high-quality grapes. The introduction of nutrients through the root or the leaves in separate phenophases of the grapevine's development ensures the normal development of the metabolic processes in the individual organs, the formation of cells and tissues, the development of the photosynthesis process, and the turnover of organic matter.

In the nutrition of the grapevine, macro and microelements should be represented in an optimal amount, otherwise, disorders occur that negatively affect the development and fruiting of the grapevine. Separately, each nutrient element has a certain specific influence on the life activity of the organs of the grapevine. Macronutrient requirements are variable in different phenophases of plant development.

Microelements have a very important role in the biochemical processes that take place in plant cells, they participate in oxidation-reduction processes, they are part of enzymes, vitamins and hormones, and they influence the synthesis of carbohydrates and their migration from the leaves to the growing parts as well as to the fruiting organs. They actively participate in many complex processes in the plant: photosynthesis, respiration as well as protein synthesis. They influence the migration and redistribution of mineral elements in the plant.

Microelements influence the intensification of the accumulation of mineral substances, their migration and distribution in the individual organs of the grapevine, which depends on the degree of maturity of the clusters and other organs. They cause an exuberant growth of the clusters in length and thickness. They affect the increase of the leaf surface, especially the young leaves that develop quickly. In this way, the general habit of the grapevine is increased, and the assimilation surface is one of the main factors for increasing the yield and obtaining a better quality of the grapes.

They influence an increase in the amount of sugars in the grape seeds, and later in the must. Participating in the physiological processes, microelements affect the increase in the concentration of cell juice and the amount of dry matter in the grapevine, thus increasing the resistance of the grapevines as well as the resistance of the clusters to low winter temperatures.

Deficiency or excess of nutritional elements affects the quality and quantity of the grapevines, and thus the wine produced.

The quality of the grapes, the colour of the berries, the taste, the size and the firmness among others are correlated with the optimal doses of macro and microelements applied during the grapevine vegetation. Correct and well-balanced nutrition has a favourable effect on the quality of grapes, while unilateral and untimely nutrition can cause undesirable chemical, physical and organoleptic changes, which deteriorates the quality of grapes.

The book *Nutrition, Quality and Biological Potential of the Grapevine* represents an innovative and unique combination of two segments of the production management of the grapevine, namely: a detailed account of the physiological role of macro and micronutrients in the grapevine and their influence on the quality and biological potential of the grape.

Namely, in the first chapter, *Physiological role, symptoms of deficiency and excess of macro and microelements in grapevine nutrition*, the nutrition of the grapevine is presented, under different environmental conditions, to improve its chemical composition and biological potential. The most important macro and microelements, their function in the grapevine and the symptoms of deficiency and excess of the nutritional elements are presented. This section also contains two appendices with the author's photos of the grapevines with a deficiency and an excess of nutrients.

In the second chapter, *Influence of macro and microelements on the quality and biological potential of grapes and wine*, the influence of nutrition with macro and microelements on the chemical composition of the grapevine is shown in detail, with special reference to the presence of the most significant biologically active components in grapes. These components have great importance on the quality and biological potential of the wine and all the products that can be obtained from the grapevine. Only with proper management of production techniques, among which nutrition is the most important, can the content of all nutritional and bioactive compounds in grapes and grape products be increased naturally, and therefore they could be classified in the functional food category.

Accordingly, this book will be of great importance among the scientific community, as it offers innovative results and presentations, but at the same time offers practical solutions that are significant for practice for the modern cultivation of grapevine.

This book is suitable for all types of readers: from researchers, scientists and professors, to students, and employees in the food industry as well as the wineries. With that, all readers will be offered a new horizon of perception for the correct and modern way of using macro and microelements in grapevine nutrition in the direction of improving the quality and biological potential of the grapevine.

Marina T. Stojanova
Dragutin A. Djukic
Monika Stojanova
Aziz Şatana

Chapter 1

Physiological Role, Symptoms of Deficiency and Excess of Macro and Microelements in Grapevine Nutrition

1. Basic Explanation

Viticulture and winemaking are important agricultural activities worldwide, but also activities that are intertwined with the lifestyle, culture and tradition of certain areas. Viticulture and winemaking, although slower than other agricultural activities are still followed the science and technology.

When it comes to winemaking (which is always associated with viticulture), the net production can vary from year to year. The conditions of the soil and climatic conditions in the production of grapes are significantly reflected in the quality of the temperature. One year the wine may be of excellent quality, but the next year the grapes from the vineyard may turn out to be of average quality.

Many factors, notably climate, soil, water and vineyard management, can influence the growth and yield components in the vineyards and therefore on grape and wine composition and quality. Vineyards are subjected to a large number of management practices including row orientation, density, pruning, clipping, tilling, soil surface management or irrigation among others, which lead to changes in the microclimate of the cluster and therefore affect grape composition (Vilanova et al., 2019).

The grapevine (*Vitis vinifera* L.) is one of the oldest cultivated plants with exceptional economic importance in Central European countries (Zebec et al., 2021).

Viticulture is a very intensive branch of agriculture that includes the cultivation of vines and the production of grapes for fresh consumption, for processing into wine, juices and other grape products (vinegar, wine distillate, etc.). Grapes and grape products are the subject of much research, especially due to the presence of biologically active secondary metabolites.

The advantage of the development of viticulture production is that it can be developed in areas where there are no conditions to grow other crops.

Vineyards are subjected to a large number of management practices including row orientation, density, pruning, clipping, tilling, soil surface management or irrigation among others, which lead to changes in the microclimate of the cluster and therefore affect grape composition (Vilanova et al., 2019).

By applying modern agrotechnical measures, with a special focus on nutrition, i.e., by applying optimal doses and types of fertilizers in an appropriate period, it is possible to significantly influence the development of the grapevine, and thus the production of high-quality grapes. To achieve complete success in viticultural production, it is necessary to know well the structure of the grapevine, its life cycle, the way of nutrition and the individual nutritional elements that are needed for proper growth, development and fruiting.

2. Historical Development of the Grapevine

The grapevine is one of the oldest cultivated plants. According to archaeological findings, man cultivated grapevines and produced grapes 8,000 years before our era. Wild forms of the grapevine are older than man. This is confirmed by fossil remains from ancient geological periods. The first fossil lineage, similar to today's, was found in the fossils of the Paleozoic period. The introduction of vines into culture first began in the regions between the Black, Caspian and Mediterranean Seas (7,000-9,000 years ago, and possibly much earlier). From there the line spread in three directions: eastward to India, southward to Palestine and Egypt, westward to the Balkan Peninsula and beyond. There are many theories about where the first grapevine cultivation began (Korać et al., 2016). The Thracians and Greeks were the first to start growing grapevines on the Balkan Peninsula. The grapevine was brought from Asia Minor through the ports of the Black Sea and the valley of the river Maritza. It is believed that from there it spread to Central Europe. The Romans were good growers and winemakers. They left a rich literature in this area. In the first century of our era, the Roman emperor Domitian, to preserve the vineyards of the Apennine Peninsula, ordered the removal of half of the vineyards in France and Spain, and in other parts of the Roman Empire, he prohibited the planting of vineyards. In the 3rd century AD, the Roman emperor Probus (276–282) lifted the order banning the cultivation of grapevines and thus enabled the expansion of the areas under grapevines in Gaul and Spain.

In the Middle Ages, viticulture was highly developed, especially in monasteries. Christianity as a religion positively influenced the spread of viticulture due to the use of wine in religious ceremonies. During Turkish rule, the development of viticulture declined due to the ban on wine production. However, the Turks from Asia Minor brought to the Balkan Peninsula some table grape varieties that are still cultivated: *Afus-ali, Chaush, Oxeye* and others. European viticulture had its greatest prosperity in the 17th, 18th and first half of the 19th centuries. Since the middle of the 19th century, its development has been threatened by diseases and pests that were not previously present in Europe.

Much research related to DNA profiling of known varieties is making progress in determining both the geographic origin and the relationship between the varieties themselves, to clarify which variety preceded which by chance or by deliberate crossing (Terral et al., 2010; Cindrić et al., al., 2019). Through cultivation, the wild grapevine (*Vitis vinifera* L. ssp. *sylvestris* (Gmelin)) gradually lost some of its morphological characteristics and gained new ones, so that thousands of years later it formed a new species - the domestic grapevine (*Vitis vinifera* L. ssp. *sativa*) which we know today. The grapevine belongs to the family *Vitacea* and the genus *Vitis* L., which includes about 70 different species, distributed in Asia, North America and Europe under subtropical, Mediterranean and continental-temperate climatic conditions (Terral et al., 2010).

Vitis vinifera is today the most widespread species of the genus *Vitis* L. It originates from the Caspian Sea and Asia Minor, where wild specimens belonging to this genus can still be found (Johnson, 1989; Bombardelli et al., 1995).

The cultivation of European grape varieties on other continents began first in South America (Peru) in the middle of the 16th century, then in North America, first in the eastern and later in the western parts, in the middle of the 17th century in South Africa, and at the beginning of the XIX century begins to be cultivated in Australia. It should be borne in mind that varieties of indigenous species were grown on these continents much earlier.

Because of its modest requirements and adaptability, the grapevine has a very wide distribution range: in the Northern Hemisphere south of 52 degrees, and in the Southern Hemisphere north of 45 degrees.

Today, grapevines are grown mostly in warm regions, primarily in the Mediterranean basin of southern Europe (Greece, Italy, France, Spain) and central Asia (Bombardelli et al., 1995). However, the vineyards of the Republic of South Africa, the United States (California), Argentina, Chile,

Australia and New Zealand are also of increasing importance. These are regions with climatic and pedological characteristics that are extremely favourable for grapevines: long, hot, dry summers and mild winters prevail, while the soil in these regions is sandy, drained and disturbed (Bombardelli et al., 1995).

3. Significance of the Vineyard Growing

Grapevine (*Vitis vinifera* L.) is one of the most economically important productive drought stress-adapted crops grown globally (Gambetta et al., 2020). Grapes and wine are widely consumed in the world, yet their mineral content can be influenced by many factors such as the mineral composition of soils, viticulture practices and environmental conditions (Daccak et al., 2022).

The significance of viticulture, among other things, is that vines can be successfully grown on soils where other crops cannot be successfully grown. These are hilly, sloping terrains, and sandy and rocky soils. Viticulture is an intensive branch of agriculture.

Great importance is also attached to the integral and organic concept in viticulture production and their development is encouraged in regions where there is no intensive agriculture and the application of large amounts of pesticides and fertilizers, that is, in conditions of uncontaminated nature.

The grapevine can be grown in backyards as an ornamental plant and at the same time a useful plant, then in small areas, for individual producers to meet their own needs for fresh grapes or wine, to supply the local market, or in larger areas with to supply the larger markets and even for export.

The most important production and economic characteristics of viticulture are:

- high consumption of human labour per unit area;
- high investment per unit area and a long period of activation (the grapevine gives the first crop in the third year) and efficiency.

Viticulture is characterized by high production values per unit area, and the local climatic and pedological conditions have a great influence, especially on the quality of the grapes. Grapevine production has a great economic value:

- grapes are important in the diet of the human population due to their high content of carbohydrates, vitamins, minerals and other nutrients;
- considering the high birth rate, grape production is a very important source of income for the agricultural population in certain areas;
- grapes are the basic raw material for the production of various types of wine, brandy and other products in the wine industry. It can be used to produce raisins, juices, concentrates and various other non-alcoholic products.

Grapes are considered one of the most popular and favourite fruits in the world for their excellent flavour, taste, and nutritional value.

Used fresh as fruit or in the form of juice, wine, raisins and jam, grapes are among the highest-quality fruits for consumption around the world. Grapes also have their uses as a raw material in the production of seed oil, anthocyanins, tannins, ethanol, etc. The use of grapes produced in different countries of the world varies significantly depending on environmental conditions. Grapes are rich in vitamins (A, B, C, E, and P), minerals, carbohydrates, lipids, proteins and polyphenolic compounds that are distributed among the skin, seeds and pulp. Grapes consist of sugar in the form of monosaccharides (glucose and fructose) and various phenolic compounds (Kandylis et al., 2021; Li et al., 2020). Grapes contain various nitrogen compounds which are grouped into two forms: mineral (NH_4^+, NO_3^-, and NO_2^-) and organic (free amino acids, nucleic acids, proteins, ethyl carbamate, and urea) (Conde et al., 2007). It also contains minerals such as potassium, magnesium, iron, and calcium as well as vitamins C, B12, and B6, which are nutritionally important (Conde et al., 2007; Pezzuto, 2008). Which metabolites will be present in the grapes and in what quantity depends mostly on the grape variety, agro-climatic factors, the maturity of the grapes during the harvest, the composition and fertility of the soil and the exposure of the grapevine to pathogens (James et al., 2022).

The yeast assimilable amino acid nitrogen is distributed in different parts in the berry: 10–15% in the seed, 19–29% in the skin, and 61–65% in the pulp (Assimilable nitrogen content must provide a good estimation of the grapevine nitrogen status (Bruwer, 2018)).

All ingredients in grapes are extremely important in the diet of the human population.

According to importance, another important ingredient of grapes is organic acids. With over 90% of must and wine, tartaric and malic acids are represented, and citric, amber, oxalic, gluconic, glucuronic and other acids are also present.

The richness of the mineral complex of grapes for the human body represents a significant source of K, Ca, Na, Mg, P, S, Fe, Cu, Mn, Al, B, I and other necessary elements. Among other ingredients, grapes contain nitrogenous substances, aromatic substances, coloured substances, tannins, polyphenols, etc. Medicinal properties of substances from the group of polyphenols (resveratrols), which are found in the seeds and skin, mostly in the black varieties, have been proven.

Polyphenols are antioxidants that eliminate harmful substances from the human body, help the body in the use of vitamin C, act preventively against heart attacks, slow down the ageing of cells, and have antibacterial and antiviral effects. In the world, especially in France, the use of grape polyphenols is highly commercialized and can be sold as dietary products in the form of capsules or the form of cosmetic preparations. The medicinal properties of grapes, grape leaves and grape brandy have long been known in folk medicine (Stojanova, 2023).

Types of grape products:

- over 80% of the total production of grapes in the world is processed into wine and alcohol;
- about 12% is used fresh as table grapes;
- about 5% is used for the production of raisins;
- other products: juice, jam, marmalade, jelly, concentrate;
- by-products: animal feed, compost, oil, tannin, tartaric acid, and coloured substances.

Ripe, cut grapevines can be used as firewood or as raw material for compost. Grapes and grape products are the subject of much modern research, especially because of the biologically active secondary metabolites. Of great importance are the studies of phenolic compounds due to their antimicrobial, antimutagenic, anticancer and antiallergic effects. According to the representation in grapes, phenolic compounds are in third place after carbohydrates and acids.

3.1. Significance of Fertilization in Viticulture Production

In modern viticultural production, one of the most important agrotechnical measures, which together with others should enable continuous, high and profitable production of grapes and their processing, is fertilizing (Stojanova et al., 2016; Stojanova et al., 2023c). The yield and quality of the grape depend on the biological properties of the grapevine, the favourable climatic conditions, the soil as well as the correct diet.

Achieving a balance between productivity and fruit quality is a major goal in viticulture, which is particularly a challenge considering the threats of climate change and weather variability. Among other factors; soil nutrient availability plays a major role in grapes productivity and quality, which creates the need for the use of fertilizers (James et al., 2022).

Also, fertilizing improves the physico-chemical and biological properties of the soil. Vineyard N fertilization has been traditionally carried out by adding nutrients to the soil so that they are absorbed by grapevine roots. Fertilizing is imposed as a necessary measure since every year, yield, and green and ripened mass of grapevine plants take large quantities of mineral substances (Gutiérrez et al., 2022).

Vegetative growth and yield of grapevine need suitable nutrition, as well as the yield and properties of grape berries, are influenced by the element deficiency so in this situation, the grapevine growers require the application of fertilizer to complement the nutrient deficiency. Balanced nutrition for grapevine improves the fruit set, fruit quality, and yield. However, improper nutrients in vineyards cause low-grade TSS, colour and maintaining quality, etc. Consequently, to obtain the preferable growth and production of grapevines, it is essential to observe nutrients on a systematic basis (Mustafa & Mustafa, 2023).

Achieving a balance between productivity and fruit quality is a major goal in viticulture, which is particularly a challenge considering the threats of climate change and weather variability. Among other factors; soil nutrient availability plays a major role in grapes productivity and quality, which creates the need for the use of fertilizers. Nutrients play a role in grapevine physiological growth, yield, and quality (Brunetto et al., 2015; Karagiannidis et al., 2007; White & Brown, 2010).

Nutrients occur naturally in the soil and are taken up by the plant roots. Too high level of nutrients causes vigorous growth of the grapevine producing a dense canopy that excessively shades the fruit or shows a varied range of toxicity. On the other hand, the soil nutrients are depleted over time and need to be replenished with organic and/or mineral fertilizers to improve grape yield and quality. Consequently, the common symptom linked to lack of nutrients is chlorosis, such that the colour of leaves changes from yellow to purple, hence restricted photosynthesis and grapes struggle to ripen properly, while quality and yield are reduced. Moreover, inadequate soil nutrients may lead to increased susceptibility of the grapevines to pests and/or diseases, which demand a large quantity of pesticides to be applied to manage them. The problem can be addressed by using appropriate fertilizers, particularly a mixed formulation that provides macro and micronutrients (Gur et al., 2022).

Fertilizer applications have a direct or long-term impact on the development, and fruiting of the grapevine and the quality of the raw material.

Managing the nutritional requirements of the vineyard requires a visual assessment of the vines, growth habits, and the nutrient status of plant tissues and/or soil to develop an appropriate fertilizer program (Arrobas et al., 2014). Each vineyard may have a unique combination of soil type, grapevine age, canopy architecture, and cultivar, thus nutritional requirements vary among vineyards, even locations within a vineyard. Analysis of the nutrient content of the petioles gives a good indication of available nutrients to the plant (James et al., 2022).

The nutrient requirements of a grapevine depend on its age, cultivar variety, yield, soil type, and properties (Holzapfel & Treeby, 2007).

Nutritional factors may have an impact on variables associated with varietal typicity; however, little work has focused on the impact of grapevine nutrition on concentrations of aroma compounds in grapes (Vilanova et al., 2019).

Vineyard nutrient deficiency can be improved by supplementation. However, high nutrient concentrations can be toxic and produce an adverse effect on grapevine performance, grape quality and their derived products (Serrano et al., 2017). Grapevines require the supplementary application of fertilizer to ensure maximum production and sustainable grape quality.

Fertilization is known to be one of the most important vineyard inputs that affect vineyard productivity and grape and wine quality (Lasa et al., 2012).

Fertilizing has a versatile impact on several physiological processes that affect the growth and development of the vegetative and generative organs of the grapevine. The vital activity of the organs of the grapevine is greater at relatively higher amounts of mineral substances.

The needs of the grapevine for nutritional elements are different in different phases of vegetative development, so nutrition should be correlated with the phenophases of development (Figure 1).

Adequate nutrient supply is important especially in the early growing season to provide the necessary structural components for bud bloom as well as successful fertilization of flowers. At the beginning of the season, due to the increase in growth in buds and the predominance of vegetative growth, the concentration of nutrients in plant tissues decreases (Keller, 2015). In these conditions, elements such as potassium due to high need in the grapevine or elements such as calcium are not provided enough for the grapevine due to the slow movement from the soil to the roots and transport in the vessels. Therefore, these conditions are considered an important factor limiting flowering, fruit set, and final fruit quality in the vineyards (Karimi, 2021).

With the correct regulation of mineral nutrition, conditions are created in which a high vegetative and reproductive potential of the clusters can be maintained, which, on the other hand, is the basis for achieving high yields and good quality of grapes.

The key to the long-term viability of a vineyard is ensuring nutrient inputs at least balance nutrient outputs via the harvested fruit, prunings (if removed), cover crop (particularly if grazed by sheep) and losses through erosion, nitrate leaching, denitrification, volatilisation of ammonia or the slow conversion of soluble P to insoluble forms (Proffitt & Campbell-Clause, 2012).

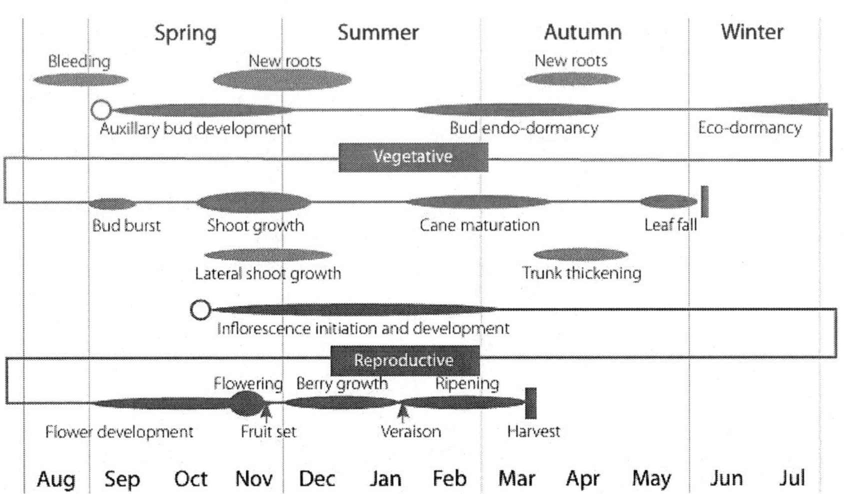

Figure 1. Seasonal growth phases of the grapevine (Proffitt & Campbell-Clause, 2012).

Grapevines require the supplementary application of fertilizer to ensure maximum production and sustainable grape quality. The use of inorganic fertilizers provides better results because of the application of known and exact amounts of fertilizers to the grapevine (James et al., 2022).

Apart from soil fertilizing, foliar fertilizing and the introduction of physiologically active substances through the leaf has a great influence on the production of high-quality grapes.

The elements needed from the soil are often inaccessible to plants. Intake of nutrients through roots requires the mobilization daily of inaccessible forms of elements in soil into accessible ones, in quantities sufficient to provide the plant with the daily nutrient quantity needed. Accessibility of elements in the soil is influenced by cold weather, soil type, pH reaction of the soil, as well as water content. Foliar feeding gives the producer assurance that the plant will absorb nutrients within a short period and be provided with them in the exact development pheno-phase when they are needed most (Duletic & Mijovic, 2014).

Deficiencies of nutrition elements can be significantly reduced by foliar nutrients of the grapevine. Feeding of grapevine via leaf is done with liquid, easily soluble fertilizers, and intended exclusively for supplemental foliar feeding of grapevine with macro and microelements, primarily with nutrients such as boron (B), zinc (Zn) and iron (Fe) which deficiency mostly occur in viticulture soils (Duletic & Mijovic, 2014).

Due to pollution troubles, however, arising from excessive use of N fertilizers in soils, foliar fertilization has once again become popular today. Despite the fact that this technique does not replace soil fertilization, it may improve the uptake and the efficiency of the nutrients applied to the plants. The processes by which a nutrient solution is applied to the leaf and is assimilated by plants include contact, leaf adsorption, cuticular or stomatal penetration, uptake and absorption into metabolically active cell compartments in the leaf, and finally, the translocation and utilisation of the nutrients absorbed by the plant (Gutiérrez-Gamboa, 2022).

Nutrient absorption by plant surfaces occurs via the cuticle, cuticular stacks, imperfections, stomata, trichomes and lenticels (Fernandez et al., 2013). The aerial parts of grapevines are covered by a hydrophobic layer called the cuticle that controls gas exchange (Müller & Riederer, 2005).

During foliar fertilizing, in certain concentrations, the grapevine is supplied with sufficient amounts of macro and microelements in an active form, which stimulates the hydrolytic processes in the tissues of the reproductive organs. During the foliar fertilizing of the grapevine, the photosynthetic processes are accelerated, which creates an additional flow of nutrients and their migration to the organs of growth and fruiting (Stojanova, 2020). The content of macro and microelements in the leaves is variable in different grapevine varieties (Table 1 and Table 2).

Foliar nutrient application to grapevines gives advantages for economic and environmental reasons. It can improve nutrient uptake efficiency as nutrients are directly absorbed by leaves. Foliar spraying provides specific nutrients to the grapevine on a timely basis during the critical stages of growth like budding, flowering and fruiting (Gautier et al., 2018; Masi & Boselli, 2011; Perez et al., 2017). A well-planned grapevine foliar nutrition program can supplement soil fertilizers especially when the grapevine root system is unable to keep up with crop demand or when soil nutrients are unavailable. For instance, fertigation, furrow or broadcasting are subjected to loss to the environment via leaching, denitrification, surface runoff, gaseous emission and microbial consumption.

Table 1. Macronutrient content (% d.m.) in leaves of some grape varieties (Stojanova et al., 2005)

Grapevine variety	N	P_2O_5	K_2O	Ca	Mg	S
Chardonnay	1.95	0.53	1.28	5.56	0.66	0.096
Italian Riesling	1.82	0.39	1.12	4.33	0.57	0.084
Cardinal	2.10	0.32	0.94	2.20	0.46	0.072
Afus ali	2.50	0.34	0.80	3.30	0.40	0.090
Merlot	1.75	0.40	1.10	3.80	0.60	0.080
Cabernet Sauvignon	1.90	0.48	0.75	4.10	0.52	0.068

Table 2. Content of microelements ($mg \cdot kg^{-1}$ d.m.) in leaves of some grapevine varieties (Stojanova et al., 2005)

Grapevine variety	Fe	Mn	Cu	B	Zn	Mo
Chardonnay	237.15	172.00	25.09	39.90	14.51	11.20
Italian Riesling	206.00	139.30	26.45	39.59	17.05	12.35
Cardinal	125.40	129.64	35.20	31.55	31.42	13.55
Afus ali	186.20	157.61	33.21	28.30	45.20	11.00
Merlot	215.30	140.20	24.90	31.50	23.60	15.10
Cabernet Sauvignon	210.20	160.00	32.85	41.90	28.10	12.25

Table 3. Visual deficiencies in grapevines
(Proffitt & Campbell-Clause, 2012)

Elements	Visual deficiencies
Nitrogen (N)	Overall reduction in growth. Leaves become uniformly light green or yellow.
Phosphorus (P)	Vines may have stunted clusters and fruitfulness is likely to be poor. Early in the season, there is a bronze/red colouration between the main veins in older leaves.
Potassium (K)	It starts as yellowing (white varieties) or bronze-reddening (red varieties) of older leaf margins. As the deficiency worsens, leaf margins become necrotic and curl upwards and interveinal chlorosis develops. Berry set can be poor.
Calcium (Ca)	Cluster tips become stunted and may die.
Magnesium (Mg)	Bright yellow (white varieties) or red (red varieties) wedge-shaped areas extend inwards between the veins on older leaves. When severe, necrosis extends inwards from the leaf margins.
Sulphur (S)	Symptoms are often similar to N deficiency (i.e., leaves become uniformly yellow, including the veins). Rare in vineyards where S sprays are used.
Copper (Cu)	Short internodes and cluster tips often die. Leaves are likely to be small, yellow and distorted. Rare in vineyards where Cu sprays are used.
Boron (B)	Cluster tip death and short inter-nodes, result in clusters with a zigzag appearance. Yellow mottling between the veins of older leaves. Edges of leaves may have small-red-brown spots. Fruit set is often poor and bunches often have 'hen and chicken' berries.
Zinc (Zn)	Short internodes, resulting in clusters with a zigzag appearance. Cluster tips have upward-curling leaves. Mottled, light-coloured interveinal colouring on leaves. Small, poorly developed bunches with 'hen and chicken' berries.
Iron (Fe)	Young leaves show interveinal chlorosis. When severe, leaves are likely to be very pale with necrotic blotches. Clusters are likely to be stunted in their growth.
Manganese (Mn)	Older leaves have a yellow mottle colour between the veins.
Molybdenum (Mo)	Necrosis of leaf margins. Poor fruit set and stunted cluster growth.

Through leaves, plants can also absorb different nutrients and export them within the stem via phloem or xylem (Dhaliwal et al., 2021), which implies that the foliar spray of micronutrients would, theoretically, determine how an applied nutrient can be reallocated from leaves to the growing tissues (Khoshgoftarmanesh et al., 2010). Foliar fertilization is one of the main methods used to improve the berry yield and quality during grape cultivation. Foliar application is the most rapid and effective method for satisfying the specific nutritional needs of plants (Fernandez et al., 2009).

The formation of the elements in the fertility of the buds is directly dependent on both soil and foliar fertilizing. Buds are the most fertile with optimal representation of macro and microelements. Under the influence of the nutritional components in mineral fertilizers, the grapes ripen faster, the bunches are more developed, and the berries are larger, with a beautiful, characteristic colour, which tolerates transpiration better. Optimum soil and foliar nutrition also affect the improvement of the chemical composition of the must. The use of macro and microelements in the nutrition of the grapevine and their participation in the oxidation-reduction processes affects the increase in the content of carbohydrates, and the decrease in the amount of total acids in the must. The content of dry matter, extracts, coloured, aromatic and mineral matters, as well as the favourable ratio of carbohydrates and acids in grapes and their processing is directly dependent on the correct and controlled soil and foliar fertilizing.

Soil and foliar fertilizing has a positive effect on the timely maturation of the grapevine clusters, which also allows for an increase in the buds resistance to low winter temperatures. There is a positive correlation between the maturation of the clusters and their degree of resistance to low temperatures, which is why mineral nutrition should enable timely maturation of the clusters and preparation of the grapevines for winter dormancy. Hence, in the ripening phenophase, it is very important that the nutritional elements, especially P, Ca, Zn, and B, are available for the nutrition of the vines (Table 3).

Macroelements in combination with microelements boron, copper and zinc affect the oxidation-reduction and a series of other processes that take place in the grapevine, which enables better maturation of the clusters and preparation of the oaks for frost resistance. The content of macro and microelements in the clusters is variable among different grapevine varieties (Table 4 and Table 5).

Managing the nutritional requirements of the vineyard requires a visual assessment of the vines, growth habits, and the nutrient status of plant tissues and/or soil to develop an appropriate fertilizer program (Arrobas et al., 2014). Each vineyard may have a unique combination of soil type, grapevine age, canopy architecture, and cultivar, thus nutritional requirements vary among vineyards, even locations within a vineyard. Analysis of the nutrient content of the petioles gives a good indication of available nutrients to the plant.

Whereas, soil analysis gives nutrient composition, which may not necessarily be available for uptake by the grapevine. Additionally, soil physical properties such as soil texture and structure may also be assessed as they influence nutrient availability. Collectively, sandy soils are likely to be leached of nutrients, and high clay content soils will rapidly fix applied potassium fertilizers. The soil high in organic content has high levels of readily available nutrients. High microbial biomass, organic carbon, exchangeable K, Ca and Mg, increased P and N availability, a balanced pH, and improved yield were reported in the vineyard treated with compost (Ball et al., 2020; Wilson et al., 2021).

Table 4. Content of macroelements (% d.m.) in clusters of some grape varieties (Stojanova et al., 2005)

Grapevine variety	N	P_2O_5	K_2O	Ca	Mg	S
Chardonnay	0.77	0.37	0.76	1.21	0.26	0.071
Italian Riesling	0.71	0.28	0.63	1.12	0.26	0.064
Cardinal	1.40	0.15	0.47	0.94	0.12	0.055
Afus ali	1.35	0.12	0.60	1.70	0.19	0.060
Merlot	1.15	0.25	0.55	1.40	0.35	0.050
Cabernet Sauvignon	0.95	0.31	0.70	1.58	0.20	0.068

Table 5. Content of microelements ($mg \cdot kg^{-1}$ d.m.) in clusters of some grapevine varieties (Stojanova et al., 2005)

Grapevine variety	Fe	Mn	Cu	B	Zn	Mo
Chardonnay	120.00	44.37	10.02	19.67	10.24	8.70
Italian Riesling	121.25	52.70	8.60	19.81	8.52	7.10
Cardinal	68.30	42.90	4.50	15.30	11.20	5.40
Afus ali	84.70	54.30	4.12	17.55	14.90	4.80
Merlot	110.20	45.95	9.50	18.55	9.20	10.20
Cabernet Sauvignon	95.30	42.30	10.80	17.10	11.55	9.60

To achieve full effects, i.e., increase the yield, improve the quality of the grapes and increase the resistance to low temperatures, soil and foliar fertilizing should be carried out in optimal amounts, aligned with the fertility of the soil, the amount of nutrient elements with the yield and the grapevine variety grown in certain climatic and environmental conditions.

4. Utilization, Role and Significance of Nutrient Elements in Grapevine Nutrition

The introduction of nutrients through the root or the leaves in separate phenophases of the grapevine's development ensures the normal development of the metabolic processes in the individual organs, the formation of cells and tissues, the development of the photosynthesis process, and the turnover of organic matter. Through regular fertilizing, the nutrients that are taken out are compensated, not only with the yield or harvest of the grapes but also with all the biomass that is rejected during the year with the pruning of the ripe or the pruning of the green. It is also necessary to provide reserves of all those nutrients that are missing in the soil to ensure proper nutrition of the grapevines in the following year (Stojanova, 2018).

The deficiency or excess of nutritional elements affects the quality and quantity of the vines, and thus the wine produced.

The most important nutrients for the vines are nitrogen (N), potassium (K) and phosphorus (P). Other elements include Carbon (C), Chlorine (Cl), sulfur (S), magnesium (Mg), boron (B), zinc (Zn), manganese (Mn), copper (Cu), iron (Fe), silicon (Si), calcium (Ca), and molybdenum (Mo), which play the role in grapevine physiological growth, yield, and quality (Brunetto et al., 2015; Karagiannidis et al., 2007; White & Brown, 2010). Their subsequent effects on must fermentation and wine quality have been identified, giving even greater importance to vineyard nutritional management.

Based on the amount of presence in the soil, as well as the function they have, nutrients are divided into two groups:

- macroelements (N, P, K, Ca, Mg, S, C, H, O);
- microelements (Fe, Mn, Cu, B, Zn, Co, Mo).

Grapevine uses macronutrients in larger quantities and they play a primary role in nutrition. Microelements are needed in a smaller quantity, but they also have great importance in the nutrition of the grapevine. Part of them the grapevine takes from nature, and part through fertilizers.

Carbon (C), oxygen (O) and hydrogen (H) are adopted by the grapevine from nature, they make up over 90% of the grapevine, and the rest are biogenic elements. The elements Na, Cl, and Si, which belong to the group of useful elements, are not necessary in the diet but are useful, and have a great influence on the development of the grapevine.

In the soil solution, the elements are found in the form of anions and cations.

Anions: HCO_3^-, NO_3^-, NO_2^-, SO_4^{2-}, $H_2PO_4^-$, Cl^-, BO_3^{3-}, MnO_4^-.

Kations: Ca^{2+}, Mg^{2+}, K^+, Na^+, NH_4^+, H^+, Al^{3+}, Fe^{3+}, Fe^{2+}, Mn^{2+}, Zn^{2+}, Co^{2+}, Mo^{4+}.

The supply of nutrients to the grapevine varies depending on:

- soil fertility;
- the activity of soil microorganisms;

- soil moisture and temperature (in wet and warm soil conditions, the root absorbs larger amounts of nutrients than in dry and cold soil conditions);
- the technologies of grapevine cultivation, etc.

The intensity of adoption of the elements is influenced by:

- the biological characteristics of the grapevine variety and rootstock (lush varieties and rootstocks with a more developed stem and root absorb larger amounts of nutrients compared to varieties with weaker and moderate luxuriance);
- the age and lushness of the plant (young plants during intensive growth and development absorb larger amounts of nitrogen compared to older grapevines in the phase of full and reduced fertility);
- the phenophase of development (the majority of nitrogen, phosphorus and potassium in the leaves of the grapevine is represented in the phase of flowering and fertilizing of the grapevine, it decreases in the phase of ripening of the grapes, and it is the lowest during the fall of the leaves);
- the ripening time of the grapes.

The whole process of the mineral nutrition of the grapevine manifests itself and acts in complete interdependence with other processes and the external environment, and it is similar to the action of individual nutritional elements. In addition to the common action on the plant, the elements also have some specific properties and activities, without the knowledge of which their useful application in modern grapevine cultivation is not possible.

4.1. How Does the Grapevine Take Up Nutrients?

The grapevine, like another plant, assimilates nutrients through the root and the leaf. It uses $CO_{2(g)}$ and $O_{2(g)}$ from the air, and the soil, it adopts mineral elements in ionic form, dissolved in water.

The fertility of the soil is important because it determines how much nutrient is potentially available to grapevines. Grapevines cultivated for wine production in general do not benefit from large amounts of water or nutrient input because grapevines are likely to exploit these resources to produce excessive vegetative growth to the detriment of wine quality. Since soil fertility is spatially variable laterally and vertically, knowledge of the variation in soil characteristics is important as this will influence the nature of grapevine growth, and water and fertilizer requirements across a particular vineyard. Soil texture is a measure of the relative amounts of sand, silt and clay particles. Texture is an important component because it determines the amount of water a soil can hold when fully wet, and the rate of water and dissolved nutrients potentially available for grapevine uptake (Rawson, 2002).

A key chemical component of soil nutrient availability to grapevines is pH. Each nutrient responds differently to changes in soil pH, with the optimum range (measured in water) for nutrient uptake being between 5.5 and 8. Soils with a pH > 8 (alkaline) can cause the following nutrients to become poorly available for grapevine uptake: P, Fe, Mn, Zn, Cu and B because they form insoluble compounds; and Ca and Mg because Na takes their place on the surface of colloids. Alkaline soils may therefore be sodic due to the abundance of Na and may also produce ammonia gas as a result of volatilisation of ammonium nitrogen. Soils with a pH < 5.5 (acidic) can cause the following nutrients to become poorly available for grapevine uptake: P and Mo because they form insoluble compounds; and Ca and Mg because they are displaced by Al and H.

In strongly acidic soils (pH < 5), Al and Mg may become freely available to grapevines at toxic levels. Soil acidity can also increase the uptake of heavy metals such as Cu and lead (Pb) and decrease the population of microorganisms (Proffitt & Campbell-Clause, 2012). The optimum pH range (measured in water) for nutrient uptake is between 5.5 and 8 (Figure 2).

Grapevine leaves have a greater number of stomata on their abaxial (the underside of the leaf) side compared to the adaxial (topside of the leaf) side, so it is a hypostomatic species. In addition, the stomata density of the grapevine leaves varies greatly in response to soil temperature and atmospheric CO_2 concentration (Rogiers et al., 2011). The abaxial side of leaves presents more polar channels than the adaxial side and this is likely the reason why the abaxial side of the leaf may absorb more N than the adaxial side (Gutierrez, 2019). The two environmental factors that most directly affect the uptake of foliar nutrient applications are temperature and pH (Fernandez et al., 2013). The increase in temperature under field conditions may increase the diffusion of solutes and solubility of the active ingredients and adjuvants but will decrease viscosity, surface tension, and the point of deliquescence of the solutions (Ramsey et al., 2005; Fernandez et al., 2013).

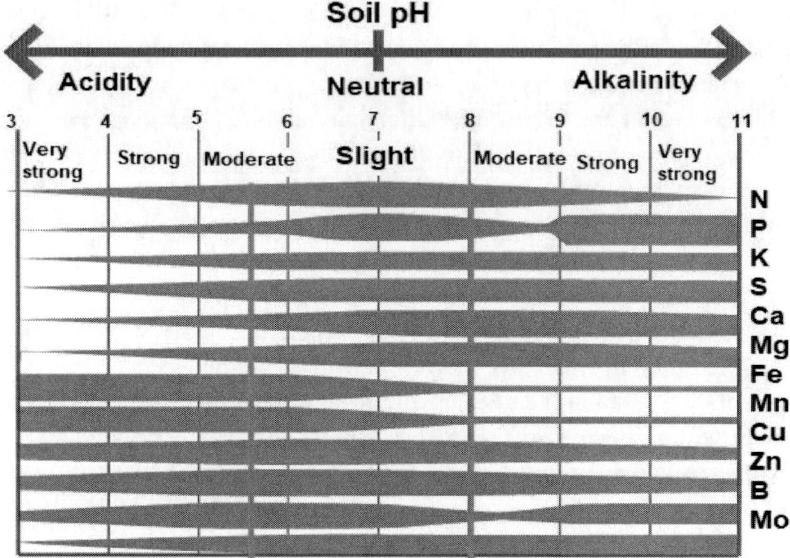

Figure 2. The effect of soil pH on the availability of nutrients to grapevines (Longbottom, 2009).

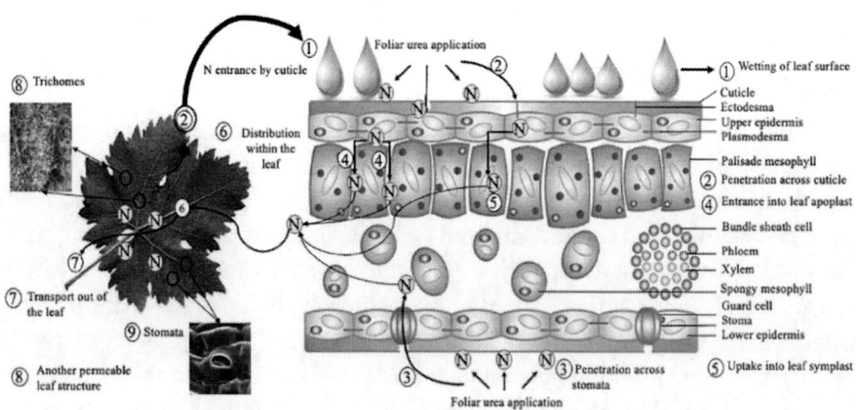

Figure 3. Nitrogen (N) entry mechanisms by foliar application to grapevines through cuticle, stomata and trichomes. Different steps of N uptake by leaves are presented. These following steps could be summarised as follows: (1) wetting of the leaf surface with N solution; (2) penetration of N across the outer epidermal cell wall; (3) penetration of N across stomata; (4) entrance of N into the leaf apoplast; (5) uptake of N into the leaf symplast; (6) distribution of N within the leaf; and (7) transport of N out of the leaf (Gutierrez-Gamboa et al., 2022).

Table 6. The major nutrients required for grapevine vegetative and reproductive growth (Proffitt & Campbell-Clause, 2012)

Elements	Major nutrients required
Nitrogen (N)	It is required for most metabolic functions and is a component of most compounds made up of and synthesised by the grapevine. Influences shoot growth in spring and prevent premature leaf fall in autumn. Influences inflorescence initiation, and berry growth and development after fruit set.
Phosphorus (P)	It is required as a component of cell membranes and genetic material and for carbon dioxide fixation, sugar metabolism, and energy storage and transfer. Influences cluster growth in spring, inflorescence initiation and fruit set.
Potassium (K)	It is required as a component of cell vacuoles and for protein synthesis and stomatal functioning. Influences shoot growth in spring, and berry size and ripening.
Calcium (Ca)	It is required as a component of cell membranes and cell wall structure and for enzymatic processes. Influences physiological disorders such as bunch-stem necrosis and the skin strength of berries.
Magnesium (Mg)	It is required as a component of chlorophyll molecules and for metabolic processes. Influences fruit formation, berry ripening and the germination of seeds.
Sulphur (S)	It is required as a component of amino acids, proteins, vitamins, enzymes and chlorophyll molecules.
Copper (Cu)	It is required as a component of oxidation enzymes and for chlorophyll synthesis and lignin formation. Influences cane maturation.
Boron (B)	It is required for the production of growth hormones, movement of sugars, pollen germination pollen tube growth and general metabolic processes. Influences cane maturation and fruit set.
Zinc (Zn)	It is required for cell metabolism, chloroplast development, hormone synthesis and pollination. Influences fruit set and intermodal elongation.
Iron (Fe)	It is required for chlorophyll synthesis and photosynthetic and respiratory processes. Prevents premature leaf fall in autumn.
Molybdenum (Mo)	It is required in converting nitrates for protein synthesis and flower functionality. Influences fruit set and is required by nitrogen-fixing bacteria.

The high temperature will speed up the rate of evaporation from the spray solutions deposited onto the leaves, thus reducing the time until solution dryness occurs when leaf penetration can no longer occur (Fernandez et al., 2013). Relative humidity may improve uptake efficacy by increasing cuticle hydration and delaying droplet drying (Ramsey et al., 2005).

Porro et al., (2010) reported that the leaf was the main organ of N recovered after foliar N application with a value higher than 60% of the total N recovered, while Verdenal et al., (2015) showed that bunches were the strongest N sink after the foliar application of urea at flowering and veraison. Such differences in N partitioning may be related to variations in leaf area and crop load of the grapevines (Figure 3).

For the normal functioning of all life functions in the grapevine, a sufficient amount of water is constantly needed. A large number of hormones and enzymes affect the regulation of the process of assimilation of nutrients from the soil into the plant, which enables and accelerates the uptake of nutrients. The mechanism of some nutrients in the plant and their transfer takes place with the consumption of energy, which originates from the products of photosynthesis and the decomposition of organic substances (Table 6).

To develop a suitable nutrition program on an individual block basis, growers need to approach nutrition holistically using the latest technology and suitable fertilizer products, with a focus on improving or maintaining soil health so grapevines can access the majority of their nutrient requirements (Proffitt & Campbell-Clause, 2012).

4.2. Movement of Nutrients in the Grapevine

All nutrients are transported through the root hairs, ie the absorption zone, to the place where they are incorporated into the individual organs of the grapevine. Some of the nutrients come quickly from the root to the point of consumption, and others very slowly. According to the speed of mobility, nitrogen is one of the fastest elements that has the greatest impact on growth. That is why this element is mostly represented in the young parts that are on the rise. According to the speed of mobility, potassium, magnesium and phosphorus follow, while some microelements (iron) are very poorly mobile and in the phase of rapid growth of the clusters during June, they cannot follow the movement of nitrogen.

The symptoms of deficiency of the individual elements depend on the speed of movement of the elements from the root to the young growing parts. If nitrogen or magnesium is deficient in the soil or if there is a blockage in their uptake, the young parts will use the nitrogen from the old leaves, so the deficiency symptoms will appear on the old leaves. If there is a lack of iron, which is poorly mobile, it is a sign that it cannot follow nitrogen in the rapid growth, so the deficiency is manifested by the appearance of chlorosis on the young parts of the grapevine. An exception to this rule is the boron, whose symptoms can be found evenly on both old and young leaves, as well as grains (Stojanova, 2023). Therefore, the term of fertilizer for each element must be adapted to the needs of the grapevine in a certain phenophase of development, as well as to the type of soil on which the vineyard was raised (Figure 4).

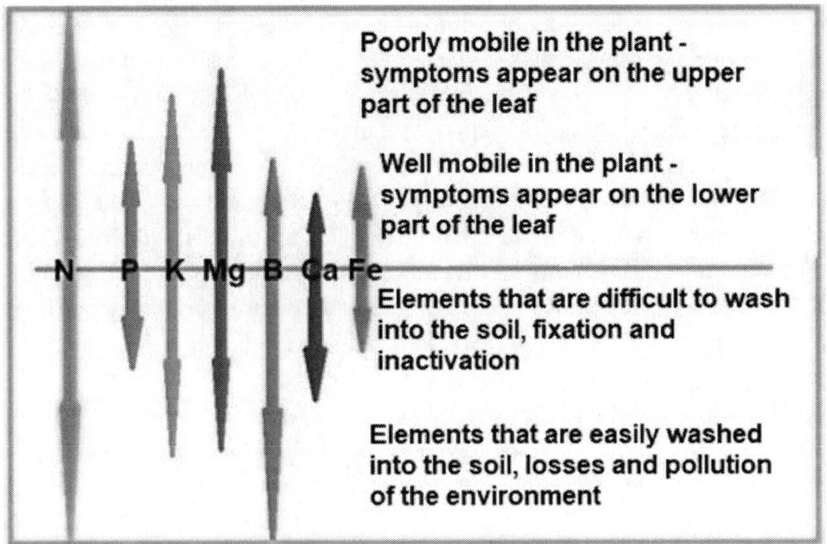

Figure 4. Speed of movement of nutrients in the grapevine and their leaching into the soil (Stojanova, 2023).

4.3. Factors Affecting the Accumulation of Ions

The accumulation of ions is influenced by many factors that can be: *endogenous* (internal) or *exogenous* (external). In natural conditions, plants are rarely exposed to the individual action of some of these factors, because they usually act on them integrally. The division exists only for easier understanding of the action of various factors and their most important effects will be briefly stated.

4.3.1. Endogenous Factors

One of the basic functions of the root is the accumulation of mineral substances, so the surface and characteristics of the root (morphological characteristics, structure and development of the root and root hairs), as well as its physiological activity mostly affect the provision of the plant with mineral elements. It is influenced by various abiotic and biotic factors, as well as the properties of the soil and applied agricultural techniques. Under favourable conditions, plants can increase the absorptive surface of the root up to 10 times. The active surface of the root consists of a zone of small thin root hairs and its size is very significant for the accumulation of elements. In search of nutrients, the root elongates and grows in the direction where they are available. The activity of the root and root hairs is significant for the accumulation of elements. The root can affect the change in the pH value of the rhizosphere (up to two pH units), and that change occurs due to the release of H^+ ions, HSO_3^-, carbohydrates, and organic and amino acids. The hairy roots formed affect the accumulation and mobilization of elements and microorganisms present in the rhizosphere. The upper part of the root secretes a mucous substance (hydrocarbon elements) which prevents the drying of the tips of the root and enables better contact and penetration of the root into the soil. of All this facilitates and increases the accumulation of substances from the soil (Stojanova, 2018).

Soil texture and organic matter affect the N retention properties, and clay soils may have higher N retention than sandy soils since they have high silt and clay, which can result in a lower cation exchange capacity (Witheetrirong et al., 2011). Silt, clay, and organic matter allow the soil to retain more NO_3 than the soils with less silt and clay (Witheetrirong et al., 2011; Stefanello et al., 2019; Gutiérrez Gamboa et al., 2020). Sandy soils normally have a low to medium level of organic matter which gives them a low capacity for supplying mineral N to grapevines (Brunetto et al., 2007; Stefanello et al., 2020a). Soil texture also affects water permeability, percolation rates and NO_3 leaching into groundwater which is faster in sandy soils (Lorensini et al., 2012; Gutiérrez-Gamboa et al., 2020; Stefanello et al., 2020a).

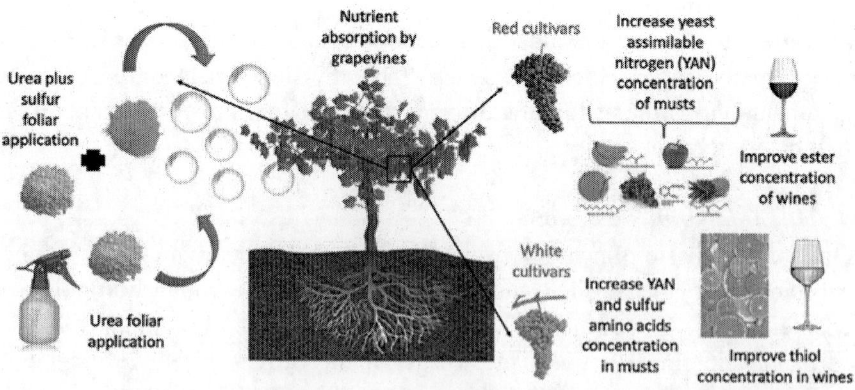

Figure 5. Summary of the effect of the foliar application of urea and urea plus sulfur to white and red grapevines with low nitrogen status on grape and wine composition (Gutierrez-Gamboa et al., 2022).

Vineyards located in tropical and subtropical regions, where rainfall is more frequent, or even at higher air temperatures, N fertilizer management should be planned carefully to favour the reduction of losses due to NO_3 leaching and NH_3 volatilisation (Lorensini et al., 2012; Brunetto et al., 2018; Stefanello et al., 2020a).

The accumulation of elements depends on the stage of ontogenesis of the plant. Maximum synthesis of organic compounds is achieved in the later stage of development. At the end of the vegetation, the accumulation of ions decreases and the concentration of certain elements in the leaves decreases because they are moved to the reserve organs, therefore in the phase of physiological maturity, the accumulation of elements is minimal. In the case of perennial crops, the accumulation of elements is stopped during the dormant period. There is a difference in the dynamism of the accumulation of certain elements during ontogenesis. So, for example, nitrogen is mostly accumulated in the vegetative phase (phenophase of main growth), because then the needs for nitrogen and protein synthesis are the greatest. Phosphorus in the accumulation has two maxima. The first appears during the period of root formation, and the other during the transition from the vegetative to the generative phase in the plant's development. Accumulation can also be influenced by the genetic specificity of plants and cultivar characteristics. Differences in sensitivity to the lack or excess of certain mineral elements have been determined among certain species and genotypes. On the other hand, specificities of plants for mineral nutrition can be created as a result of their adaptation to certain environmental conditions, their morphological and anatomical features, or their metabolism.

There are also indicator plants that accumulate specific ions. More recently, plants have been selected that have a greater resistance to an excess or lack of an element or a lower ability to accumulate some elements (especially heavy metals). Plant health can affect element uptake. Various biotic factors (pathogens, harmful insects, etc.) can lead to the occurrence of diseases in plants and reduce not only the uptake of mineral substances but also their assimilation. The effect is especially harmful if the disease occurs in the root (Stojanova, 2020).

Endogenous factors also affect the foliar uptake of ions. Genotypic characteristics, varietal differences, leaf traits (such as cuticle thickness), leaf age (young leaves absorb more than old leaves), nutritional status of the plant, and ontogeny stage (accumulation is greater in intensive growth of the plant).

In summary, the advantages of the foliar application of urea to grapevines are that the application costs are lower than those of soil fertilization, the uptake of the nutrients is not dependent on the roots, and a low quantity of fertilization is needed to supply N needs (Figure 5). Foliar N fertilization is also preferred over soil N fertilization where the topsoil is dry, the soil has low nutrient availability, and the grapevine root activity is decreasing or in vineyards cultivated on sandy soils (Gutiérrez-Gamboa et al., 2020). Some authors reported that foliar application of urea to grapevines increased the concentration of several amino acids in must compared to the control (Garde Cerdan et al., 2014; Verdenal et al., 2015; Hannam et al., 2016).

4.3.2. Exogenous Factors

Of the external factors that influence the adoption of the elements, the properties of the soil are particularly significant. The concentration of available forms of mineral elements depends on them. The uptake of elements depends on the chemical and physical properties of the soil. The pH value of the soil directly affects root growth (optimal pH value is 5.5-6.5) and the accumulation of ions. At higher concentrations of the pH value of the soil, cations are mainly accumulated, and at lower concentrations of the pH value, anions are accumulated. The pH value indirectly affects the availability of elements in the soil, the activity of microorganisms in the soil (in soil with a low pH value, fungi dominate, and in soil with a higher pH value, bacteria dominate) and, in particular, on the accumulation of ions. Soils with an acidic pH value (4.5-5.5) have less carbonates and more silicates and are rich in heavy metals and phosphates. The availability of some elements, such as Zn, Cu, Mn, P and Mg, is higher in moderately acidic soils than in alkaline soils. Acidophilic plant species are, for example, the sweet chestnut or the blueberry. Alkaline soils (pH 6.5-7.5) are rich in carbonates and poor in heavy metals and phosphates. In alkaline soils, the availability of Mo is higher than in moderately acidic soils.

For the physiological processes of the root to take place, the presence of oxygen in the gaseous phase in the soil is necessary, that is, the soil should be aerated. A low oxygen concentration below 10% causes a reduced accumulation of ions, which usually occurs in compacted-type soils or soils with excess water (saturated with water). The osmotic pressure of the solution also affects the uptake and accumulation of ions. The low osmotic potential of the external solution or soil can lead to reduced accumulation and transport of elements. The exception is certain ecological groups of plants (halophytes) that thrive in saline soils, where the values of the osmotic potential of the soil are low (Stojanova, 2018).

Microorganisms in the soil are an important factor in fertility and help in the uptake of elements. Of special importance is the symbiotic community of the hyphae of the fungus (about 60 species) with the roots of higher (multicellular) plants (mycorrhiza). Mycorrhizae can be facultative or obligate. It is present in dicotyledons (83%), monocotyledons (79%), and gymnosperms. Mycorrhiza is not represented in representatives of the family *Cyperaceae, Brassicaceae*, in aquatic plants, as well as in dry and salty soils.

Mycorrhiza can be represented as ectomycorrhiza or endomycorrhiza. Ectomycorrhiza occurs in forest trees (pine, spruce, oak), where mycelium densely envelops the tip of the root, and free hyphae emerge from the soil as root hairs. Endomycorrhiza occurs when the hyphae of the fungus penetrate the interior of the root cell and into the intercellular spaces where they branch. At the same time, the external hyphae engulf the soil and thus increase the absorption surface of the root from 100 to 1000 times. Such symbiotic relationships allow the plant to improve the supply of water and mineral substances, and increase the accumulation of nitrogen and phosphorus, while the fungi make humus more available and increase the resistance of plants to drought, pathogens and heavy metals. Fungi in their cell walls bind metals in complexes and thus reduce their toxic effect. For fungi, mycorrhiza is significant because they receive carbohydrates from the host plant, which are necessary for their metabolic processes.

The concentration of the ions is the factor that most affects the uptake, because as the concentration increases, so does the uptake of the ions. This indicates that plants are not completely selective and may take in more elements than they need. However, uptake is not linear but decreases at high ion concentrations.

The absorption of mineral elements is influenced by various other factors, such as temperature, light, amount of water in the soil, air humidity, etc. An increase in temperature directly affects the increased permeability of the cell membrane of the root and leads to a rapid uptake of elements. The indirect influence of temperature on ion uptake is seen through the effect of photosynthesis, transpiration and respiration processes (Stojanova, 2020).

The water regime of the soil affects the solubility and concentration of elements in the soil, as well as the transpiration process.

All this proves that for the optimal nutrition of plants, it is not enough only to provide the soil with mineral elements that are necessary for their growth and development, but also the factors affecting the availability of elements and their uptake by plants must be taken into account. It is of particular importance for the cultivation of plants in which the optimally implemented agrotechnical measures (fertilizing, irrigation, tillage, etc.) and the selection of the appropriate genotypes, can enable the maximum use of the available elements in the soil or the substrate in which it is grown and nurtures the plant and achieves a good and quality yield.

Most of the nutritional needs of trees are added to the soil in the form of organic and chemical fertilizers, but due to barriers to solubility and ion uptake, only a small part of these elements is transferred to the aerial parts of the tree. These limitations can affect negatively both the growth and development of fruits especially when there is a greater need for nutrients. At the beginning of the growing season, due to increased growth in buds and the predominance of vegetative growth, the concentration of nutrients decreases dramatically in grapevine tissues (Karimi et al., 2020; Keller, 2015).

4.4. Balance and Antagonism of Individual Nutritional Elements

In the nutrition of the grapevine, all macro and microelements are necessary and equally important. For proper nutrition of the grapevine, it is necessary to have a balanced ratio of nutritional elements.

Individual nutrients are represented in different amounts both in the leaf and in other organs of the plant.

With foliar analysis, results are obtained for the content of nutritional elements, which determines the deficiency, surplus or toxicity of individual elements in the grapevine.

Therefore, determining the necessary and optimal ratios of the most important nutritional elements is one of the most complex problems in grapevine nutrition. In its organs, the grapevine accumulates nutrients for the next year, which is not the case with annual crops.

Table 7. Causal-consequential reactions between antagonistic pairs (Stojanova, 2020)

Reason	Consequence
Low K content	Increased N uptake
High K content	Decreased absorption of Mg
Low N content	Increased P uptake
Low S content	Decreased absorption of N
High Ca content	Decreased absorption of Fe and P
High Mn content	Decreased absorption of Fe and Mg
High Al content	Decreased absorption of P

Several nutrient ions are known to be mutually antagonistic, meaning that high amounts of some nutrients in the soil can block the absorption of other elements. Such is the case with K and Mg, Mg and Fe, K and NH_4^+ etc.

The mutual ratios of the ions in the soil affect their uptake and assimilation by the plants, on the content in the plants as well as on the quantity of yields. Those relationships can be *antagonistic* and *synergistic*. These ratios in nature usually occur on extreme soils and when large doses of fertilizers are applied.

The reasons for the appearance of antagonism between ions are different. They can occur due to differences in ion diameter, valence or charge differences. The occurrence of antagonism can occur as a result of the mutual action of individual ions that create difficult-to-access compounds for plants in the external environment or in the plant itself. Ion antagonism can also occur when a greater amount of one ion in the soil solution prevents the uptake of another ion. For example, a low content of K^+ ions in the soil affects an increased uptake of nitrogen, while a reduced content of nitrogen stimulates the uptake of phosphorus. On the other hand, for nitrogen to be properly absorbed, there must be a certain amount of sulfur in the soil. The reduced amount of available nitrogen affects the increased uptake of phosphorus (Table 7).

The antagonism of the ions is explained by their competence in the process of adoption. Therefore, the ratio of available nutrients in the soil is of great importance for their adoption and the impact on the quality and quantity of production.

When the mentioned elements are present in the grapevine, different characteristics are observed about their presence in the soil.

To correctly determine the deficit or surplus of an element, and thus determine the appropriate types and doses of fertilizers, among other things, the antagonism between individual elements must be taken into account.

4.4.1. Examples of Antagonism

A high amount of nitrogen in the soil affects the reduction of the absorption of phosphorus and potassium and reduces the content of manganese.

The presence of optimal amounts of phosphorus in the soil removes the toxic effect of aluminium and excess molybdenum. Adding large amounts of phosphorus to the soil reduces the absorption of zinc and magnesium.

Table 8. Antagonistic pairs of elements (Stojanova, 2020)

NH_4^+–K	K–B	Mn–Mo	Zn–Fe
NH_4^+–Ca	P–Fe	Mn–Zn	Ni–Fe
NH_4^+–Mg	P–Zn	Cu–Mn	Cr–Fe
K–Mg	P–Al	Cu–Fe	Co–Fe
K–Ca	Mn–Mg	Cu–Mo	SO_4^{2-}–Mo
K–Na	Mn–Fe	Cu–Zn	

At a higher amount of calcium, the absorption of potassium is reduced. If there is magnesium in the soil in an optimal amount, it positively affects the absorption of phosphorus and reduces the absorption of potassium and calcium. But if there is less of it in the soil, the absorption of phosphorus and iron decreases (Table 8).

The antagonism between copper and iron is manifested by the increased occurrence of chlorosis. Excess copper reduces the iron content of chloroplasts.

An excess of zinc affects the reduction of absorption of iron and manganese.

The antagonistic effect of iron is expressed in manganese, nickel and cobalt.

The antagonism of boron with copper, manganese and iron is most pronounced at the beginning of the vegetation when the development of the vegetative mass is the greatest so that the plants in that period absorb the mentioned nutrients the most.

Ion synergism is a phenomenon in which the uptake of one ion is stimulated by the uptake of another ion. Examples are frequent when anions stimulate the uptake of cations and vice versa.

5. Physiological Role of Macroelements

A large part of the macroelements are necessary to carry out complex biochemical and physiological processes and the grapevine adopts them from the external environment. The macro-nutrients, including nitrogen (N), potassium (K), and magnesium (Mg), are essential elements that impact grapevine growth, grape yield and wine quality. Currently, plant tissue or soil sampling, followed up with chemical analysis, informs on whether a nutrient is deficient, adequate, luxurious or in toxic concentrations (Proffitt, 2012; Zlámalová et al., 2016). These nutrients, in particular N and K have high mobility and can be translocated rapidly from older to new leaves if the demand in young leaves or fruiting bodies is higher than the supply from the soil, resulting in chlorosis and necrosis (Jegadeeswari et al., 2020; Michopoulos & Solomou, 2019). In the nutrition of the grapevine, macro elements should be represented in an optimal amount, otherwise, disorders occur that negatively affect the development and fruiting of the grapevine. Separately, each nutrient element has a certain specific influence on the life activity of the organs of the grapevine. Macronutrient requirements are variable in different phenophases of plant development.

5.1. Carbon (C), Oxygen (O) and Hydrogen (H)

These three elements form the basis of every living organism. At the same time, carbohydrates are created from them, which further play an important role in the photosynthesis process. The source of these constituent elements is air and water. Carbon in the air is in the form of CO_2 and is represented by 0.03%. Oxygen in the air is present in an amount of 21%. The plant uses oxygen as needed, it enters the composition of organic compounds and is consumed during their decomposition, during which energy is released. Hydrogen becomes available to the plant through the process of photolysis. Carbon, oxygen and hydrogen are represented in nature in sufficient quantity for the needs of plants. However, with timely and correct tillage of the soil, by improving the water-air regime, it is possible to influence their increased representation, thereby creating favourable conditions for their adoption through the root.

5.2. Nitrogen (N)

Nitrogen (N) is an essential soil-derived macronutrient element for plant growth (James et al., 2022). The grapevine adopts nitrogen in the form of NO_3^- and NH_4^+ ions, but also in the form of organic molecules depending on their availability in the soil, as well as based on other factors. For plants, NH_4^+ or NO_3^- ions have identical significance, although the nitrate form is a more common source of nitrogen. Which of these ions will be taken up by plants depends on several factors including the pH value of the soil environment (NH_4^+ is more taken up in alkaline and neutral soil, and NO_3^- is more taken up in acidic conditions), the concentration of the solution, the presence of other ions that act antagonistically, the type of plant or climatic factors. Uptake of both ions is an active process that depends on metabolic energy.

Nitrogen, among nutrients that soil supplies to grapevines, is the most influential on grapevine physiological growth, vigour, and yield as well as grape and must quality (Arrobas et al., 2014). Nitrogen plays a key role in plant metabolism. As a macronutrient, it represents approximately 1.5% of the dry weight (% DW) of grapevine and enters the composition of key metabolites, such as proteins, amino acids (AAs), enzymes, DNA, RNA and chlorophyll (Verdenal et al., 2021).

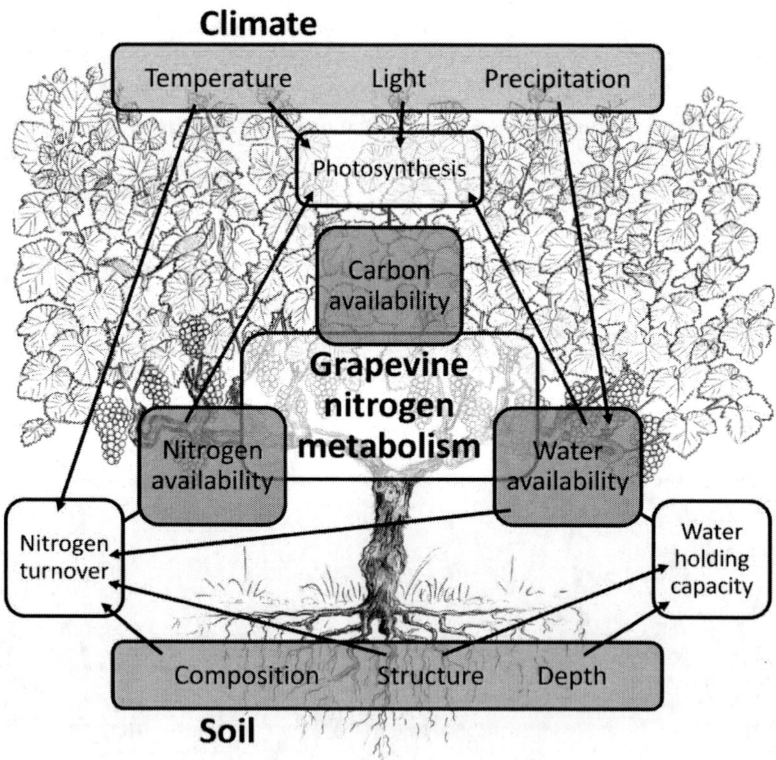

Figure 6. Impacts of environmental conditions (i.e., climate and soil) on grapevine N metabolism (Verdenal et al., 2021).

This comes from the fact that nitrogen is a constituent element of various compounds essential for plant life: a structural component of proteins, amides, amines, nucleic acids, chlorophyll, pigments, enzymes, coenzymes, nucleic acids, alkaloids, etc. Alanine, glutamic, and aspartic acid are essential amino acids that are formed by the direct coupling of an organic acid and ammonia, while others are formed by the transfer of the amino group from an amino acid to a keto acid. Glutamine and asparagine are the most important amides because they have the role of eliminating excess ammonia in plants.

The grapevine belongs to the group of nitrogen-heterophilic plants. As an integral part of both nucleic acids and chlorophyll, nitrogen has a very important role in the photosynthesis process and affects the normal metabolism of the grapevine, which results in increased fertility. It positively affects the development of vegetative organs and the assimilation surface of the leaves, increasing their productivity. It affects the content of soluble dry matter. It also affects the water regime of the plants (the greater amount of nitrogen increases the intensity of transpiration). It affects the absorption of ions (NO_3^- anion promotes the collection of cations such as potassium, calcium and magnesium, and NH_4^+ cation promotes the collection of phosphorus, chlorine and sulphur). It affects the resistance of plants to stressful factors (increased amount of nitrogen affects the reduction of resistance of plants to low and high temperatures, droughts and diseases).

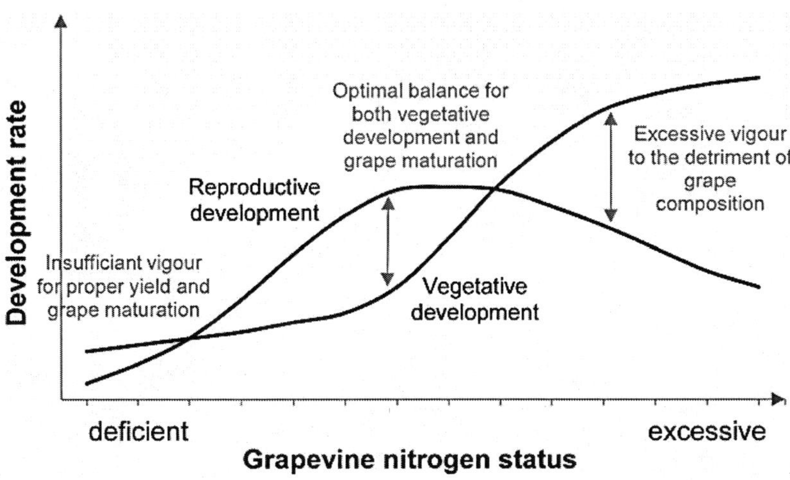

Figure 7. Hypothetical model of vegetative versus reproductive development rates as a function of grapevine N status (Verdenal et al., 2021).

Without a good nitrogen supply, good growth and grapevine productivity are not possible, because the metabolic processes of this element provide energy for all other physiological processes: photosynthesis, synthesis of growth regulators, respiration, and absorption (Figure 6).

The grapevine needs nitrogen from the very beginning of the vegetation for proper growth and development of the total green mass; the needs decrease during the reduced growth until the beginning of the ripening of the grapes, and during the ripening of the grapes and the emergence of the version, the nitrogen needs increase again. During leaf fall, nitrogen uptake does not occur. N status alters both grapevine production variables and grape composition to different degrees (Schreiner et al., 2018). Vegetative growth is more constrained than reproductive growth as N status decreases, as illustrated in Figure 7 and Figure 8.

The distribution of the adopted nitrogen in the organs of the grapevine is uneven; it is mostly in the leaves up to 50% of the total adopted mass, and then in the seeds. During periods of intensive absorption, it accumulates in the vegetative organs up to 60%, and in the generative organs up to 40% (Figure 9).

Several reports have shown that foliar application of urea to grapevines may affect basal bud fertility, stored N reserves, the amino acids concentration of grapes, chlorophyll, carotenoids, volatile and phenolic substances, including stilbenes, and grape microbiota (Verdenal et al., 2016a; Brunetto et al., 2018; Gutiérrez-Gamboa et al., 2018b).

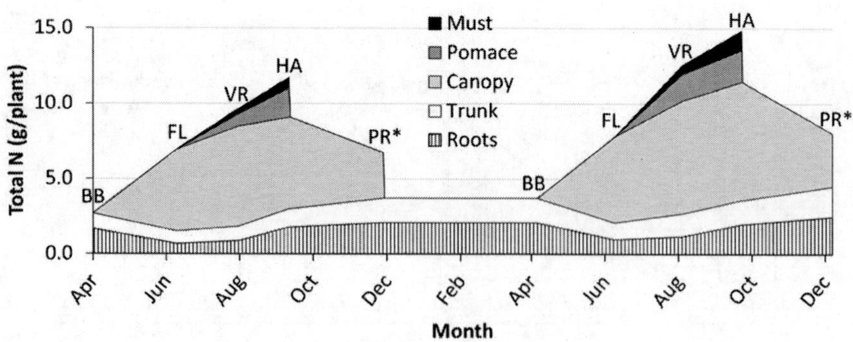

Figure 8. Changes in N content of plant parts in grapevines over two growing seasons (Verdenal et al., 2021).

It has been determined that the properties of the grapevine substrates significantly influence the absorption of nitrogen and its content in the above-ground organs of the grapevine, and the different shapes and heights of the stems, as well as the differences in the distances between the grapevines, have no effect on the absorption of nitrogen.

Mineral nitrogen from the soil, due to its rapid transformation into nitrates, can easily be leached from the soil, which is a significant environmental problem. Knowing the redistribution of nitrogen during the vegetation cycle (Figure 10) is an important segment for the correct implementation of fertilizing. From the total added nitrogen at the beginning of the vegetation, about 20% is adopted, which in the verasion phase is equally distributed between the vegetative and generative organs of the grapevine. After the verasion phase, most of the total nitrogen (45%) is redistributed to the bunches.

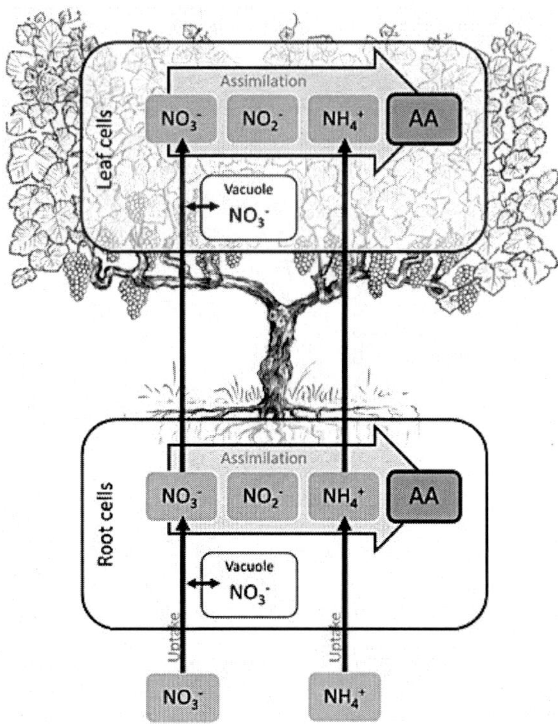

Figure 9. N uptake and assimilation in grapevine. NO_3^-, nitrate; NO_2^-, nitrite; NH_4^+, ammonium; AA, amino acid (Verdenal et al., 2021).

Apart from fertilizing, knowing the redistribution of nitrogen is important for grape processing because the amount of nitrogen coming into the must from the grapes is significant for the correct fermentation of the must. Nitrogenous compounds (amino acids) are required for the growth of yeasts during the must fermentation process. Optimal nitrogen fertilizing leads to an increase in the amount of nitrogen compounds in grapes and must. During winemaking, N in the must can nourish yeast growth and metabolism and ultimately modify the flavour components in wine, which influences the quality and sensory attribute of wine (Gong et al., 2022; Holzapfel et al., 2015; Miliordos et al., 2022).

The dynamics of N uptake and the impact on the physiology and grapevine yield are necessary for the proper application of nitrogenous fertilizers. Proper application of N fertilizers ensures not only high yield but also, balanced vegetative and reproductive growth in the plants. The increase in grape cluster weight and length as well as yield depends on the N doses (Stefanello et al., 2020).

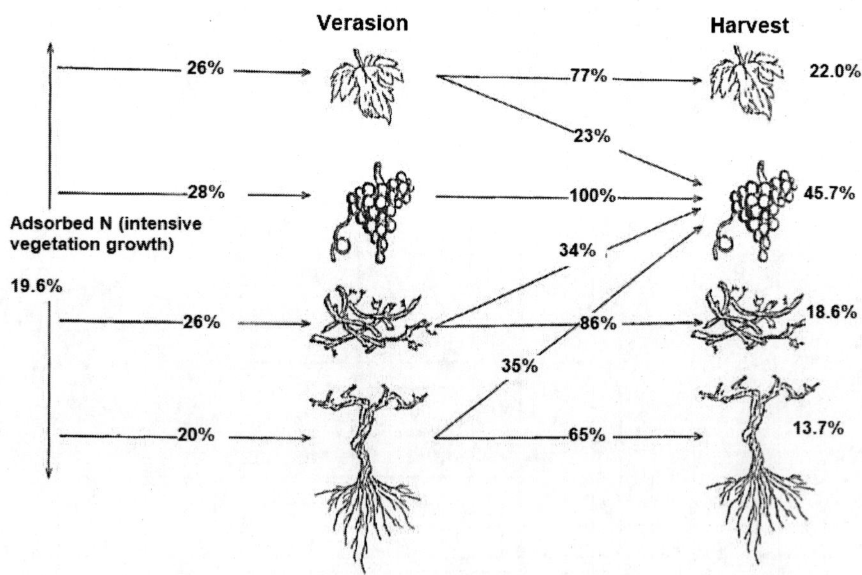

Figure 10. Nitrogen redistribution during grapevine vegetation (Stojanova, 2023).

Imbalanced use or widespread use of N fertilizers alone favours vegetative growth of lateral clusters hence dense canopy associated with reduced fruit sugar and delayed ripening, which causes poor quality of the fruits in terms of total soluble solids (TSS), colour, keeping quality, mineral content, high malic acid, low must pH, and the fruits become susceptible to sunburn (Cocco et al., 2021).

Adequate N availability is required to support optimum grape yield and fruit quality (Havlin et al., 2022).

5.2.1. Deficiency of Nitrogen

Nitrogen deficiency results in reduced vegetative growth and photosynthesis, small-thin and stiff leaves, slowed cluster growth, shortened internodes, and low berry set (Cocco et al., 2021; Verdenal et al., 2021). A deficiency of N in grapevines, however, may affect key metabolic functions and retard cluster development and bunch formation (Keller, 2020). Nitrogen deficiency is manifested by the appearance of pale green to yellow-green leaves, which simultaneously become narrower and shorter. The leaves are small and thin, and the petioles turn reddish. At the same time occurs low berry set, reduced long-term bud fruitfulness and yield (Guilpart et al., 2014), reduced grape N content and possible delayed maturation (Schreiner et al., 2018). Leaf symptoms appear after the beginning of ripening as the result of N translocation from the leaves to the berries. When N deficiency is accompanied by water stress, the leaf margins may roll slightly upwards, wilt, and dry. Sometimes tissue between the veins turns light brown and dies. Under extreme N deficiency, the leaves wilt and drop (Tian et al., 2022). N excess leads to high vigour, dense canopy, large dark-green leaves, extended vegetative growth period (competing with and delaying grape ripening) and increased grape sensitivity to fungal diseases (Thomidis et al., 2016).

In addition, unilateral nutrition with nitrogen in clusters slows down the formation of hard layers, increases the content of pith, reduces the content of dry matter and starch, slows down the maturation of the clusters and reduces their resistance to low winter temperatures. Disturbances are first noticed on the older leaves, and then they affect the younger leaves as well. In addition to slowing down photosynthesis, the yield and quality of grapes decrease. The colour of the berries is weaker. Grapes and wine have a weaker aroma.

Figure 11. Progression of N deficiency in the grapevine variety Chardonnay from left to right (Rogiers et al., 2021).

When there is a significant lack of nitrogen, the grapevine has a poor condition, that is, all organs (clusters, leaves, bunches) lag in development, and yellowing begins to occur on all leaves. In the later stages, the edges of the old leaves fade, necrotize and eventually fall (Figure 11). The size and number of bunches on the clusters decrease. The wine has poor quality, it is inharmonious and without emphasized varietal properties. During fermentation, N plays an important role as a food source for yeast growth and performance. Therefore, low grape nitrogen content impacts the fermentation rate such that lessened YAN content in the must inhibits yeast growth and metabolism, which could lead to stuck and sluggish fermentations (Schreiner et al., 2018; Tian et al., 2022). In the absence of nitrogen, the berries are very small, which significantly reduces their market value.

Inadequate grapevine nutrition further leads to varying nitrogen compound quantities in fruit, which has a determining effect on the fermentation processes. Excessively low nitrogen compound levels in grapes distort the alcoholic fermentation closing processes (Canoura et al., 2018).

5.2.2. Excess of Nitrogen

Contrary to nitrogen deficiency, excess of nitrogen also has a detrimental effect on grapevines. In the case of excessive fertilizing with nitrogen, especially in conditions without sufficient and balanced fertilizing with phosphorus and potassium, unnaturally lush growth occurs with dark green leaves and long and thick clusters from which many sprouts emerge (Schreiner et al., 2014). Although such a grapevine visually looks nourished, due to the excess of nitrogen, a spongy tissue is formed that is resistant to fungal diseases, drought during the summer months and low winter temperatures. The relationship between the vegetative and reproductive potential is disturbed, it causes the flowers to fall, fertilization also fails, and the bunches become red, which reduces the quality of the grapes and the wine.

The excess of nitrogen causes the extension of vegetation until late autumn, the clusters do not mature enough, the resistance to diseases and the resistance to low winter temperatures decreases, and the relationship between vegetative and birth potential is disturbed. In conditions of insufficient provision of the soil with potassium, the excess of nitrogen affects the intensification of the manifestation of this deficiency. Excess nitrogen also accelerates the appearance of chlorosis.

High N uptake can also increase the susceptibility of the grapevines to diseases, such as powdery and downy mildew, and bunch stem necrosis in which developing flower clusters are aborted by the grapevine causing severe yield reduction. This is in alignment with the study of Add El-Razek et al., (2011), who revealed that high N results in increased juice per berry and decreased fruit firmness.

High soil N inhibits root growth, which makes grapevines more susceptible to water stress, pests and diseases. Thus, nitrogenous fertilizers should be applied to ensure an adequate N supply during spring development, but available N in late summer should not be high enough to encourage late-season cluster growth, delayed maturing and promote immature canes (James et al., 2022).

When very high doses of nitrogen fertilizers are applied, there is a drying of the edges of the leaves which then curl downwards. As a consequence of the extremely lush growth, there is a decrease in the yield and the quality of the grapes (high content of acids and proteins). High doses of nitrogen have a detrimental effect on the quality of must and wine. Considering that excessive fertilization with nitrogen delays the ripening of grapes, the wines obtained from vineyards on soil rich in nitrogen have poorer quality, poorly expressed aroma, and are more difficult to clarify, stabilize and store (Table 9).

Excess nitrogen also has a detrimental effect on the quality properties of grapes. In such conditions, the grains usually lose their characteristic taste, their flesh becomes softer, especially in rainy years, and they are more difficult to maintain during storage. In conditions of excess nitrogen, the content of carbohydrates in the must decreases.

Table 9. Symptoms of nitrogen (N) deficiency and excess (Moyer et al., 2018; Stojanova, 2023)

Symptoms	Deficiency	Excess
	• Pale green leaves, and in severe deficiency the leaves are small; • Grapevine growth can be slow, with short internodes; • Reduction of the number of bunches and berries; • Weak pigmentation in red varieties.	• Large and dark green leaves; • Redness of the bunches • Reduced yield; • Extension of the vegetation, reduced resistance of the grapevine; • The wine is of lower quality.
Where (on the plant)	• The first symptoms appear on old leaves; • With weak deficiencies, they usually appear first on the basal or middle leaves as a fading of the green colour of the leaf, while with large deficiencies, there is a more pronounced yellowing (chlorosis) of the leaves.	• On the green vegetative parts.
Where (in the vineyard)	• In vineyards with a low content of organic matter; • On sandy and shallow soils.	• Groundwater contains a high content of nitrates; • Soils with a high content of organic matter.
When	• Deficiency symptoms can appear at any time, but are more common in spring during intensive growth (when N is mobilized); • In dry periods when N cannot be absorbed by the plant.	• During the vegetation.

5.3. Phosphorus (P_2O_5)

The grapevine adopts phosphorus from the soil solution in the oxidized form in the form of anions ($H_2PO_4^-$ and HPO_4^{2-}) of orthophosphoric acid that can dissociate in water, incorporating it into organic matter without reduction (Davenport & Jones, 2016). Phosphorus is taken up as primary and secondary phosphates depending on the pH value of the soil. In a weakly acidic environment, primary phosphates are mostly represented, and in soil with a weakly alkaline pH value, plants use secondary phosphates. Plants can use phosphorus from pyrophosphoric and metaphosphoric acid salts, whose content in the soil solution is not negligible. Phosphorus is needed evenly throughout the vegetation.

For the grapevine, phosphorus is a very important element of mineral nutrition, because many important physiological processes take place in the grapevine, and most of them are related to phosphorus metabolism. The content and distribution of phosphorus in the organs of the grapevine are closely related to the processes that take place in these organs.

In viticulture, phosphorus is called an energy element because its physiological role is related to the supply of the plant with energy compounds (ATP and ADP) that are necessary for the proper metabolism of the grapevine. Phosphorus participates in all important biochemical processes in the plant (photosynthesis, glycolysis, cellular respiration and energy transfer). It is an important element for the creation of generative organs, it favourably affects the development of flowers, the ripening of grapes and stem, the resistance to low temperatures, and thus the quality of the wine.

Phosphorus is part of many organic compounds found in the structure of protoplasm and favourably affects the exchange of substances. The specific role of phosphorus in plant metabolism is determined by the structure of its anions. Phosphate ions represent tetraoxy anions because they contain four hydrogen atoms, three of which can participate in the formation of different types of bonds (hydrogen, ionic and covalent bonds) with other molecules. That is why phosphorus is an important structural component of various organic compounds.

The absorbed phosphorus is quickly transferred to the plant and in a certain order, it is incorporated into organic compounds – phosphoric acid esters and phospholipids, which as hydrophobic and hydrophilic molecules are important in cell membranes and organelles. As an integral part of nucleic acids, phosphorus participates in the synthesis of proteins, nucleotides and many important coenzymes such as ADP.

As a component of ATP, phosphorus participates in all energy processes that take place in the cell. Phosphate esters represent the cell's reserve metabolic energy. Phosphorus is necessary for many physiological and metabolic processes, such as (Stojanova, 2023; Schreiner & Osborne, 2018; Zheng et al., 2020):

- photosynthesis and respiration (Krebs cycle, glycolysis), because these processes cannot take place without the presence of phosphate esters and phosphorylated carbohydrates;
- synthesis of starch and transport of the products of photosynthesis, because the transport of triose phosphates from the chloroplasts to the cytoplasm takes place through P/TP-transporters (translocators);

- metabolism of carbohydrates and proteins, so the lack of phosphorus increases the amount of low-molecular compounds;
- respiration, transformation of carbohydrates, and has a special significance in the creation of chlorophyll and nucleic acids. Nucleic acids and proteins represent a basic component of protoplasm, necessary for the construction of the nucleus, plastids and mitochondria;
- transport through membranes (phosphorus is a component of phospholipids);
- regulation of the pH value of the cytoplasm;
- taffects the quality and quantity of yield.

The lack of phosphorus can disturb the metabolism of plants, especially the metabolism of carbohydrates and nitrogenous compounds, and therefore its amount in the leaf must be optimal. The concentration of phosphorus in the plant tissue of the grapevine decreases with increased fertilization with nitrogen.

The total amount of phosphorus in the grapevine is more represented in the vegetative organs (56%) and less in the generative organs (44%).

Large amounts of phosphorus are found in seeds during the formation of the endosperm and embryos, in the growing points and the phloem and represent mobilization centres for providing the necessary energy.

Phosphorus is very important because it provides energy transfer in the plant, which is why it is especially important in the initial stages of growth it encourages the development of the root, strengthening the whole grapevine, thus affecting the maintenance of fitness and increasing resistance to low winter temperatures. The more developed the root system, the better the overall condition of the grapevine. Optimum nutrition of the grapevine with phosphorus has a favourable effect on the formation of flower buds, on faster ripening of the grapes and encouraging full ripening of the clusters, and also affects the increase of resistance to high and low temperatures. The intensity of phosphorus absorption is positively correlated with the growth of organs and with the dynamics of metabolic processes. During the vegetation period, the phosphorus content gradually decreases, especially after fertilizing. During grain growth, phosphorus is translocated from the leaves to the clusters.

Table 10. Symptoms of phosphorus (P_2O_5) deficiency and excess (Moyer et al., 2018; Stojanova, 2023)

Symptoms	Deficiency	Excess
	• Early season: dark green leaves. • Mid-late season: reddish-purple leaves in red varieties; light pink leaves in white varieties; • Reduced yield and quality.	• There is antagonism with Fe, Zn, Mn, etc.; • Appearance of brown spots on the leaves.
Where (on the plant)	• Symptoms are characteristic of the basal leaves.	• On the leaves.
Where (in the vineyard)	• Cold and wet soils in spring can reduce the availability of P; • At soil pH < 6.5 and > 7.5.	• In soils where high doses of phosphorus fertilizers have been applied.
When	• Symptoms are usually manifested before the beginning of vegetation.	• During the vegetation.

5.3.1. Deficiency of Phosphorus

The lack of phosphorus also occurs in conditions when it is present in the soil in sufficient quantity, but there is a lack of microelements: boron, zinc, manganese, and molybdenum, due to which the intensity of transformation of complex phosphorus compounds into simpler ones decreases and their movement decreases in the berries and the leaves. Very often, on the soils under vineyards, phosphorus is blocked and unavailable to the plants due to the extreme acidity of the soil. On such soils, it is necessary to first solve the problem of acidity, that is, to perform calcification of the soil. Under these conditions, there is a delay in the development and ripening of the grapes, as well as the appearance of symptoms characteristic of phosphorus deficiency. To have greater availability of mineral phosphorus for plant nutrition, the soil under the vineyard should be neutral, slightly acidic or slightly alkaline (Stojanova, 2023).

With a lack of phosphorus, the overall growth and development of the grapevine slows down. The first symptoms of phosphorus deficiency, as with nitrogen, first appear on the lower, oldest leaves, which are initially dark green to blue-green (due to increased chlorophyll formation) with a metallic sheen. Later, due to the increased formation of anthocyanins, purple shades appear on the edges of the leaves in red varieties, while in white varieties the colour of the leaves changes to brown-yellow. In the terminal phase, the leaves acquire a dark bronze colour and necrosis occurs (Schreiner & Osborne, 2018). Phosphorus deficiency reduces fruit set, resulting in loose and small clusters (Davenport & Jones, 2016). The woody part of the clusters is thin and weak, the root system is poorly developed, the flower buds fall off, the berries are less tightly bound, the vegetation period of the grapevine is extended, and the still green leaves fall. The yield and quality of the grapes decrease, as well as the grapevine's resistance to diseases and low winter temperatures, and the beginning of vegetation is delayed in the spring. Plants become highly susceptible to diseases and pests. With a lack of phosphorus, a lack of magnesium also occurs very often (Table 10).

5.3.2. Excess of Phosphorus

In the nutrition of grapevines, excess phosphorus is a very rare phenomenon mainly due to the low fertility of the soil with this element. With an excess of phosphorus, the chlorophyll in the leaves decreases, the ageing is accelerated, the leaves fall prematurely, the flowering is earlier and the grapes ripen faster. Excess phosphorus causes a disturbance in the absorption of microelements (Fe, Zn, Mn, etc.), due to which symptoms of their deficiencies are manifested.

5.4. Potassium (K_2O)

The grapevine very intensively absorbs potassium from the soil solution or the soil adsorptive complex in the form of K^+ ions. In the soil solution, potassium is found in the form of easily soluble potassium salts (carbonates, sulfates or chlorides), the availability of this element depends on the rate of diffusion. In the soil, potassium moves slowly from the upper to the lower layers. However, this negative property is compensated by the ability of the grapevine to create a reserve amount of potassium in the root and stem. Potassium can also be absorbed through the leaf, but due to the large needs, the needs cannot be fully met in this way. In plants, the uptake of K across the membrane is facilitated by channels and transporters either of which can be high or low affinity, and sometimes both. Given the ability of K to translocate from roots to leaves and then back to roots, the grapevine is an extremely responsive and adaptable system, capable of maintaining internal homeostasis and driving nutrient flow to areas of greatest demand (Rogiers et al., 2017). The direction of potassium transport (through the xylem or phloem) depends on the physiological needs of the plants, but also the provision of the plants with this element. Potassium is a highly mobile macronutrient that is integral to many physiological and biochemical processes within plants. K has a strong role in regulating the membrane potential of the cell and therefore is critical to the uptake of other ions and sugars. It is essential for plant signalling, osmoregulation, maintaining cation-anion balance, cytoplasmic pH regulation, enzyme activation and protein and starch synthesis (Rogiers et al., 2017).

Potassium is one of the most important elements for the quality of grapes, must and wine. It is also the most important nutritional element for regulating the metabolism of the grapevine (Figure 12).

Potassium is not a constitutive element and does not enter the structure of organic compounds, but without potassium carbohydrates and proteins cannot be synthesized, nor a large number of other syntheses can be achieved. The function in plant metabolism is the result of its influence on certain processes of creating more organic compounds, especially amino acids, which are synthesized under its influence creating proteins.

Potassium affects the intensity of photosynthesis, protein synthesis, transport and accumulation of carbohydrates, nitrogen metabolism, water regime, grapevine resistance to drought, low temperatures and diseases.

Potassium as an essential element of grapevine nutrition has an important physiological role that can be divided into two basic functions:

- enzyme activation;
- regulating the permeability of living membranes.

Potassium is an activator of enzymes necessary for photosynthesis and respiration and is very important for the osmotic potential of the cell (it has a significant role in osmotic regulation and ion balance).

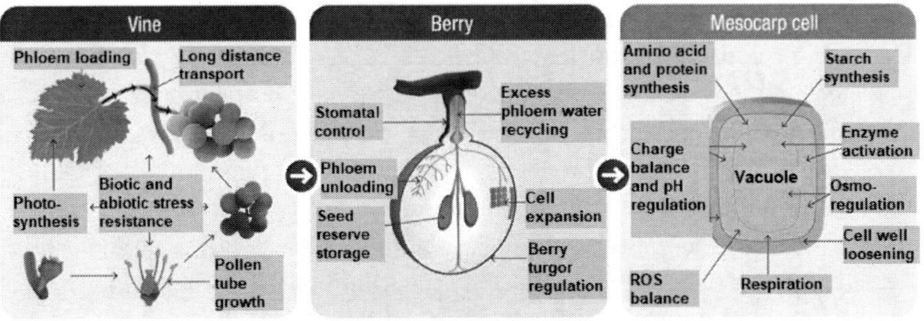

Figure 12. A speculative model describing the functions of K in the grapevine at the whole-plant, fruit and cellular level (Rogiers et al., 2017).

During the grapevine's growing season, there are four maxima in potassium requirements:

- cluster growth;
- flowering;
- the beginning of a verasion;
- physiological maturity of the grapes.

Potassium can improve the effect of transferring other elements and cause better loading of photosynthetic products such as glucose and sucrose into the phloem vessels during flowering, fertilization, and fruit set (Keller, 2015; Menegel, 2007).

The greatest amount of potassium is needed in the flowering and spreading phenophases. The increase in the concentration of potassium in the leaves has a positive effect on the greater synthesis and translocation of carbohydrates and thus encourages earlier ripening of the grapes. The optimum supply of potassium to the grapevine affects reduced water consumption for the synthesis of dry matter, that is, potassium increases the amount of dry matter in clusters, leaves and wine. It affects the increase of the content of carbohydrates and aromatic compounds in the berries, reduces the total acidity of the must and wine, and increases the resistance to drought, diseases, mechanical pressure and winter frosts, while the lack of potassium leads to a faster loss of turgor and less resistance on drought. All plants that produce starch and carbohydrates need a higher amount of potassium. For that reason, the ratio of nutrients N:P:K for grapevines should be 1:2:3.

The importance of potassium in the nutrition of the grapevine is indicated by its high content in the leaf (about 2.5%), often many times more than nitrogen, which explains its important role in photosynthetic processes. Accelerating the synthesis of ATP, and potassium, in the light phase of photosynthesis, provides the energy needed for the assimilation of CO_2 and the synthesis of organic compounds. In addition to the positive influence on the elasticity of the cell wall, the permeability of the cell membranes and the effect on the reduction of organic acids in the berries during their ripening, potassium increases the resistance of the grapevine to pests and diseases. This is achieved by the influence of potassium on changes in the concentration of cell sap, lignification of cell walls and more frequent closing of stomata.

The grapevine uses potassium evenly throughout the vegetation, and the greatest need is 20 to 35 days after flowering and during the ripening of the grapes. As the grapes ripen, organic matter from the leaves moves into the grains. Potassium stimulates this process by increasing the permeability of the cell walls - primarily in leaf stalks and bunches. The transfer of potassium from the leaves to the berries starts when they begin to soften. The process is very intensive and stops only when 75–80% of the total potassium content of the leaf migrates into the bunch.

The interaction of K with other elements is complex and the use of Ca and Mg fertilization to reduce K uptake has come up with inconsistent results (Hannan, 2011). Ca itself can have profound effects on berry development (Hocking et al., 2016). This is likely due to soil and climate interactions but does warrant further experimentation. All these approaches may also induce K deficiency and thus precise recommendations are required that are based on real-time measurements of grapevine and berry K status. Finally, little attention has been given to the role of the bunch microenvironment on K accumulation, mainly radiation, temperature, evaporative demand and wind speed. This is despite many studies confirming the role of the microenvironment on other attributes such as sugars (Dai et al., 2011), acidity (Sweetman et al., 2009), colour (Matus et al., 2009) and aromas (Azuma et al., 2012).

5.4.1. Deficiency of Potassium

The lack of potassium usually occurs on light, sandy, calcareous and clay soils, as well as in dry climatic conditions. Potassium deficiency in grapevines is most often the result of low potassium levels in the soil, an underdeveloped root system and unfavourable soil moisture. In cold and wet years, the symptoms of potassium deficiency can be noticed already in early spring.

The symptoms of potassium deficiency differ in the individual phenophases during the vegetation. In the spring, red-brown necrosis occurs on the edges of the young leaves, while necrosis rarely occurs on the old leaves during the summer. In drought conditions, glassy spots appear between the nervature of the leaves.

The lack of potassium affects the overall growth and development of plants because potassium participates in all important functions in metabolism. Due to its high mobility, it moves from the older to the younger parts, so the deficiency affects the slowing down of plant growth.

Potassium deficiency primarily manifests itself on the lower leaves, producing yellow-brown to brown spots (Figure 13).

Figure 13. Progression of K deficiency in the grapevine variety Chardonnay from left to right (Rogiers et al., 2021).

The edges of the leaves are bent down, and then their drying occurs. The time of onset of visible symptoms of potassium deficiency correlates with the degree of deficiency. It most often occurs during the phase of intensive development of the bunches, when the movement of potassium from the leaves to the berries is the most intense. In case of severe deficiency and drought conditions, the symptoms may appear earlier. With a slight deficiency, the potassium content of old leaves can ensure moderate growth of young leaves, but they remain short and, on the other hand, thin due to the limited growth of the cambium.

Table 11. Symptoms of potassium (K_2O) deficiency and excess (Moyer et al., 2018; Stojanova, 2023)

Symptoms	Deficiency	Excess
	• Weak deficiency: yellowing of the leaves that starts from the edge and progresses towards the centre; • Major disadvantage: chlorotic areas become necrotic and appear burnt; • The edges of the leaves curl, and the grapevine has reduced growth and yield; • Reduced content of soluble and coloured substances in the bunch.	• Antagonism with Mg, B, Ca, Zn and Mn.
Where (on the plant)	• On the basal leaves.	/
Where (in the vineyard)	• On light, sandy, calcareous, clay soils; • In soils that are rich in exchangeable calcium and magnesium; • Underdeveloped root system, insufficient humidity in the soil.	• In conditions of excessive fertilizing with potassium fertilizers.
When	• At the end of summer.	/

In the case of a lack of potassium, weak and reduced growth of the grapevine is observed, the root branches poorly with a decrease in the number of root hairs, which results in a decrease in the absorption of water and nutrients, which on the other hand directly affects the decrease in the resistance of the grapevine in dry conditions (Table 11).

The lack of potassium has a negative impact on the growth and ripening of clusters and grapes, on the reduction of the accumulation of carbohydrates in the grapes and the increase of the acidity of the must, and later of the wine. The sensitivity of grapes to the occurrence of *Botrytis* increases.

In the K-deficient vineyards, the supply of sink organs with photosynthates is impaired and sugar accumulates in source leaves. Limits cluster and fruit growth and results in premature leaf fall, poor fruit set, decreased fruitfulness, delayed ripening, and low yield (Griesser et al., 2017; Rogiers et al., 2017).

Also, low K supply results in low fruit K concentration as well as low must pH (Schreiner et al., 2013).

5.4.2. Excess of Potassium

Excess of potassium is more difficult to notice, except in conditions of excessive fertilization. In practice, it rarely happens because potassium moves slowly from the surface to the deeper layers of the soil. The excess of potassium negatively affects the absorption of other ions, especially in conditions where there is not enough water. An excess of potassium causes a deficiency of Mg, B, Ca, Zn and Mn because it makes their absorption difficult.

5.5. Calcium (Ca)

The grapevine absorbs calcium in ionic form (Ca^{2+}). Calcium has a significant physiological role in the life of the grapevine. It is found in free and bound form and affects cell division and growth, especially the growth and development of the root system. The presence of calcium in almost every cell organelle and cytoplasm indicates its great indirect or immediate importance for the development of metabolic and physiological processes. Its content increases in older leaves. Calcium is a constituent of a small number of organic compounds in the plant. As a divalent cation (Ca^{2+}), it is unable to form intramolecular complexes, but it can connect molecules in intermolecular complexes, and this is one of the basic meanings of its action. It affects the physical and chemical properties of protoplasm, the stability of cell membranes and the activity of enzymes.

It also affects the formation of mitochondria and germination as well as the growth of pollen tubes. The role of calcium in the plant cell is extremely important because it is an integral part of the calcium pectinate found in the cell wall and thus has a significant role in maintaining the structure of the cell walls and increasing the viscosity of the protoplasm (Duan et al., 2022).

Calcium has a great influence in regulating the permeability of the cell membrane because it has been proven that in the absence of calcium the membrane permeability increases, whereby various substances and ions enter the cell more easily and the whole process leads to negative consequences.

Calcium contributes to auxin activity and is involved in cell division and cell elongation, germination, and pollen tube growth. Calcium is effective in improving flowering, maturation, and transfer of carbohydrates from leaves to fruits. It seems that an adequate supply of nutrients during the plant growth period while increasing the effective pollination period has increased the percentage of fruit set (Karimi et al., 2021).

The physiological role of calcium is related to the processes of photosynthesis, respiration, protection of plants from the toxic effects of an excess of certain elements, reduction of hydration of protoplasm, etc.

There is an ionic antagonism of Ca^{2+} to K^+ and Mg^{2+}, which can slow down the absorption of these elements.

The harmful accumulation of calcium is greatest in the xylem. The increased targeting of calcium in the vegetative organs, especially in the clusters and leaves, contributes to the timely maturation of the clusters, which is of great importance, especially in dry years. The influence of calcium on the development of the root system of the grapevine is determined by its positive influence on improving the soil structure and encouraging the development of beneficial microflora.

5.5.1. Deficiency of Calcium

The lack of calcium in the soil or the increased concentration of other cations (H^+, K^+, Mg^{2+}, NH_4^+) can lead to a problem in supplying plants with calcium. These cations as antagonists of the Ca^{2+} cation can reduce its uptake. Also, poor mobility of calcium can lead to its deficiency (especially in generative organs). This is most pronounced in drought conditions, as well as in higher humidity in the soil. These conditions reduce transpiration flow, making it impossible for calcium to be efficiently transported to young tissues and generative organs.

Table 12. Symptoms of calcium (Ca) deficiency and excess (Moyer et al., 2018; Stojanova, 2023)

Symptoms	Deficiency	Excess
	• Necrosis of the tips of green shoots; • The leaves bend and the nervature is green; • Flowers are stunted before flowering.	• Chlorosis due to antagonism with Fe, K, Mg, B and other microelements.
Where (on the plant)	• On the tips of green shoots and leaves; • The leaves are wrinkled, the ends are bent up, the appearance of necrosis.	• On the upper parts of the young leaves and green shoots, and spreads towards the lower parts of the grapevine.
Where (in the vineyard)	• With abundant fertilizing with potassium and magnesium fertilizers.	• On carbonate soils with a high pH value.
When	• With the appearance of the leaves.	/

However, the appearance of calcium deficiency is not only related to lower amounts of this microelement in the soil but also to its insolubility, i.e., unavailability for plants. The occurrence of insoluble forms of calcium is common in soils rich in phosphorus, which forms insoluble compounds with calcium. Otherwise, the availability of calcium is also associated with abundant fertilizing with potassium and magnesium fertilizers.

Figure 14. Progression of Ca deficiency in the grapevine variety Chardonnay from left to right (Rogiers et al., 2021).

Symptoms of calcium deficiency are usually caused by an excess of K, Mg and N. The first signs of deficiency are observed at the tips of green shoots due to the low mobility of calcium in the grapevine. At the same time, the growth of the grapevine is reduced, the flowers are stunted, the leaves are wrinkled, the ends are bent upwards, in certain places necrotic tissues appear between the nerves and the leaf dies, and the whole plant lags in general development. The root remains small and rudimentary (Figure 14).

5.5.2. Excess of Calcium

Excess calcium causes chlorosis because it blocks the absorption of Fe, K, Mg, B, Mn, Zn and other trace elements. Symptoms of excess calcium are manifested on the top parts of young leaves and green shoots and spread to the lower parts of the grapevine (Table 12).

5.6. Magnesium (Mg)

Magnesium in the soil is found in the form of easily soluble salts which are available to plants as the Mg^{2+} cation. Magnesium from the soil solution is passively absorbed and transported through the root system. Unlike calcium, magnesium has greater mobility and is easily transported through the xylem and phloem. Apart from the root, it can also be absorbed through the leaf. The absorption and transport of magnesium is related to the absorption of water. Compared to calcium, magnesium, being more mobile, can be translocated from older to younger organs.

Magnesium (Mg) is an important macronutrient with several physiological functions that influence plant physiological growth and development. It is the central atom of chlorophyll and it activates enzymatic processes, thus contributing to carbohydrate production in leaves through photosynthesis. Also, favourably influences assimilation, such that, high Mg levels may limit the uptake of K by the vein (James et al., 2022).

This element has a very important role in almost all physiological processes in the plant such as photosynthesis, biosynthesis of proteins, metabolism of nucleic acids, water regime of plants, growth and development, and quality and quantity of yields.

Grapevines have greater needs for magnesium because it is a biogenic element that enters into the composition of chlorophyll, is a structural component of chlorophyll molecules (bound to the four pyrrole rings) and is necessary for its synthesis. Of the total amount of magnesium found in the grapevine, 15 to 20% is found in chlorophyll, and the rest is in the form of salts and free or bound ions in cell walls and enzymes. It is an activator of a series of photosynthesis enzymes needed for CO_2 binding. Magnesium affects the transport of the products of photosynthesis in the phloem and the storage of spare carbohydrates; it is an activator of many metabolic enzymes (dehydrogenases, carboxylases, decarboxylases); is a cofactor of almost all enzymes that catalyze substrate phosphorylation; it is necessary for nucleic acids (it is a cofactor of enzymes that catalyze hydrolysis and the creation of ester bonds that are important for the processes of transcription, translation and replication of nucleic acids); affects the structure of ribosomes and protein synthesis; it also participates in the construction of the cell membrane in the form of magnesium pectinate; affects nitrogen metabolism (magnesium deficiency increases the amount of non-protein compounds); affects the water regime (magnesium deficiency leads to an increase in stomatal openings and intensification of the transpiration process). The greatest need for magnesium begins at the stage of early grape ripening.

Throughout the grapevine, magnesium moves mostly acripetally and is therefore mostly found in the upper aerial organs. The absorption of magnesium depends on the climatic conditions, i.e., a humid climate is more favourable. In the absence of Mg, the K/Mg balance shifts in favour of potassium. The large accumulation of nitrogen, phosphorus and calcium in the leaves negatively affects the uptake of Mg and the synthesis of chlorophyll. The properties of the grapevine varieties and the grapevine rootstocks have a significant influence on magnesium absorption.

5.6.1. Deficiency of Magnesium

The excess of magnesium occurs on sandy soils, on acidic soils, then on soils with a heavy mechanical composition and on cold soils, where due to the lack of optimal temperature in the zone of the root system, there is no absorption of magnesium. Magnesium deficiency can also occur in conditions where it is present in sufficient quantities in the soil. The uptake of magnesium by plants can be significantly reduced as a result of the antagonism of magnesium ions with potassium, calcium, sodium and ammonia ions. Antagonism between magnesium and sodium ions occurs most often in the case of irrigation with water containing a high amount of sodium.

The lack of magnesium is first manifested in the photosynthesis process whose intensity is minimized due to the reduced synthesis of chlorophyll as well as the reduced activity of important enzymes. The metabolism of carbohydrates and the transport of assimilates through the phloem are also disturbed so that carbohydrates and starch accumulate in the leaves. Protein synthesis also decreases, and the amount of amides increases. The concentration of hydrogen radicals increases, which can lead to oxidative stress. Plant growth is reduced resulting in reduced yield. A lack of magnesium leads to a delay in the fermentation of the must, which produces more acetic acid and less alcohol.

Magnesium deficiency symptoms are often associated with manganese deficiency symptoms. The difference is that magnesium deficiency symptoms first appear in older leaves, while manganese deficiency is characteristic of young leaves. In extreme cases of deficiency, the leaves fall.

Magnesium deficiency symptoms appear on the oldest leaves in the form of chlorosis, just like iron deficiency. Deficiency symptoms appear in the form of yellowish and reddish spots on the leaf mesophyll between the nerves. In such conditions, the formation of chlorophyll and the synthesis of carbohydrates ceases, the clusters ripen poorly, and the grapes are of poorer quality. With a severe lack of magnesium, the leaf first turns orange, then red, then purple, and then parts of the leaf turn into necrotic surfaces, while the leaf veins are still green. With a prolonged lack of magnesium, the leaves completely dry up. Also, increases the risk of tendril atrophy. Collectively, chlorosis is caused either by Mg, N, P, Mn and/or Fe deficiency, high content of soil calcium (calcareous soils) or a combination of these factors (El-Ezz et al., 2022).

Table 13. Symptoms of magnesium (Mg) deficiency and excess (Moyer et al., 2018; Stojanova, 2023)

Symptoms	• Deficiency	Excess
	• Mild deficiency: chlorosis (yellowing) starting near the edge of the leaf and progressing inwards. The tissue immediately surrounding the nerves remains green; • Major disadvantage: chlorotic areas become necrotic; • In red varieties, the leaves acquire a reddish colour.	• Antagonism with K and Ca.
Where (on the plant)	• Appears first on the oldest leaves; • Reduced growth and yield.	• There are similar symptoms of calcium deficiency on the leaves.
Where (in the vineyard)	• In acidic soils (pH < 5.5), sandy, cold soils.	• On soils created by the decomposition of the parent-substrate dolomite.
When	• At the end of summer.	/

Table 14. Summary of nutrient disorder symptoms for well-watered Chardonnay and Shiraz grapevines grown in greenhouse conditions at 25 °C (Rogiers et al., 2021)

Nutrient disorder	First leaves to show symptoms	Necrotic leaf margins	Interveinal chlorosis	Distinct necrotic legions across blade	Leaf rolling	Vein colouration
N deficiency	Older	X	Even yellowing	X	X	Normal
P deficiency	Emerging leaves	X	X	X	Emerging leaves- downward rolling	Normal
K deficiency	Older	Yes	Striations	Yes	Downward rolling of mature leaves	Normal
Mg deficiency	Older	X	Striations	Yes	X	Normal
Ca deficiency	Younger	Yes	X	Yes	X	Normal
Fe deficiency	Younger	X	Even interveinal chlorosis across the blade	X	X	Dark green major and minor veins giving intricate pattern
B deficiency	Cluster tips necrotic	X	X	X	Cupping of young emerging leaves	Normal
B toxicity	older	Yes	Yes	Yes	Cupping of emerging leaves	Normal

Figure 15. Progression of Mg deficiency in the grapevine variety Chardonnay from left to right (Rogiers et al., 2021).

Magnesium can reutilize, that is, the ability to move from the old to the young leaves, which is why visible symptoms of deficiency appear in the lower young leaves during the vegetation (Figure 15). Due to the mobility of magnesium, the changes are characteristic of old leaves. Flowering stops, and the flowers are light in colour.

It is easily leached from the soil, so magnesium deficiency symptoms are more pronounced in wet years (Table 13 and Table 14).

5.6.2. Excess of Magnesium

Symptoms of excess magnesium are rarely seen. An excess of magnesium can occur on soils created by the decomposition of the parent substrate dolomite. Excess magnesium is toxic to plants. Magnesium is an antagonist of potassium and calcium. Based on the theory of ion antagonism, the calcium ion prevents the excessive intake of magnesium ions, thereby preventing the possible harmful effects of magnesium. It causes a lack of calcium, as a result of which the root system and above-ground organs of the plant show symptoms that resemble a lack of this element. Therefore, for proper nutrition, growth and development of plants, the correct ratio between calcium and magnesium is needed.

5.7. Sulphur (S)

Plants absorb sulphur through the root in inorganic form as sulfate anions (SO_4^{2-}), mainly sulfates of K, Na, Ca and Mg which are then reduced and used for the synthesis of organic compounds. Part of the sulphur in the form of SO_2 can be absorbed from the atmosphere or by applying mineral fertilizers and using plant protection products, which contain sulphur among other elements. Sulphur is an important nutrient required for grapevine physiological growth and development. It helps to form proteins, coenzymes and chlorophyll, plays a role in energy metabolism, and promotes root growth and seed formation (Considine & Foyer, 2015).

In the physiological processes of the grapevine, sulphur also has a significant role, although the needs for this element are smaller. The synthesis of some amino acids, proteins and vitamins cannot be achieved without its participation. Physiologically active parts of the grapevine have a high content of sulphur, and in conditions of good supply of potassium, sulphur contributes to increasing the content of carbohydrates in grapes. With most of the amount of sulphur, the grapevine is supplied by applying mineral fertilizers containing it (potassium sulfate, ammonium sulfate). In the organs of the grapevine, sulphur accumulates and is stored in the form of sulfate, and according to physiological needs, it is chemically reduced and incorporated into organic compounds.

5.7.1. Deficiency of Sulphur

Symptoms of sulphur deficiency are similar to nitrogen deficiency, yet are rare given the widely adopted use of sulfur-based sprays for the management of downy and powdery mildew and sulphur-containing fertilizers in most vineyards globally (Ferreira et al., 2022; Savocchia et al., 2011). The lack of sulphur leads to disruption of physiological processes and plant metabolism. At the same time, protein synthesis decreases, soluble amino acids (asparagine, arginine), carbohydrates and nitrates accumulate (sulphur is necessary for the activity of nitrate-reductases), and the intensity of photosynthesis also decreases. The first symptoms of sulphur deficiency appear on the youngest leaves because translocation from old to young leaves is very small. The leaves turn green-yellow and yellow, which reminds of a lack of nitrogen, but it should be noted that the symptoms of nitrogen deficiency are first noticed on the oldest leaves, and with a lack of sulfur, the symptoms are manifested in the youngest leaves, which is why it is necessary to examine the ratio between the content of nitrogen and sulphur in the leaves. Damage starts from the top of the plant (Table 15).

Table 15. Symptoms of sulphur (S) deficiency and excess (Moyer et al., 2018; Stojanova, 2023)

Symptoms	Deficiency	Excess
	• Green-yellow and yellow colour of the leaves.	• A high concentration of SO_2 in the air causes partial or complete defoliation of the stem.
Where (on the plant)	• On young leaves.	• There is damage to the plants, which first results in the appearance of chlorosis, and then necrosis, first along the edges of the leaves, and then on the entire leaf surface.
Where (in the vineyard)	• Areas with heavy rainfall (leaching of sulphur); • Arid regions.	• Rare occurrence in soil.
When	• At the end of summer.	/

5.7.2. Excess of Sulphur

Excess sulphur can occur in the soil or the atmosphere. It rarely occurs in the soil, mainly on poorly aerated soils, where in addition to the reduction of sulfates, H_2S can also accumulate, which in higher concentrations is toxic to plants.

In the atmosphere, especially in industrial zones, there can be an increased concentration of SO_2, which is easily oxidized to SO_3, and then easily dissolves with water droplets in the atmosphere, creating first sulphurous (H_2SO_3) and then sulfuric acid (H_2SO_4). Sulfuric acid reaches the Earth's surface in the form of harmful acid rain.

The harmful effects of excess SO_2 are manifested in many processes in plants: photosynthesis decreases, drought stress or biotic stress can occur, and partial or complete defoliation of the stem can occur. There is damage to the plants, which first results in the appearance of chlorosis, then necrosis, first on the edges of the leaves, and then on the entire leaf surface.

The symptoms of deficiency and excess of individual macroelements are illustrated in Appendix A (Figure 16 to 25).

Appendix A: Symptoms of Deficiency and Excess of Macroelements (Stojanova, 2023)

Figure 16. Deficiency of N.

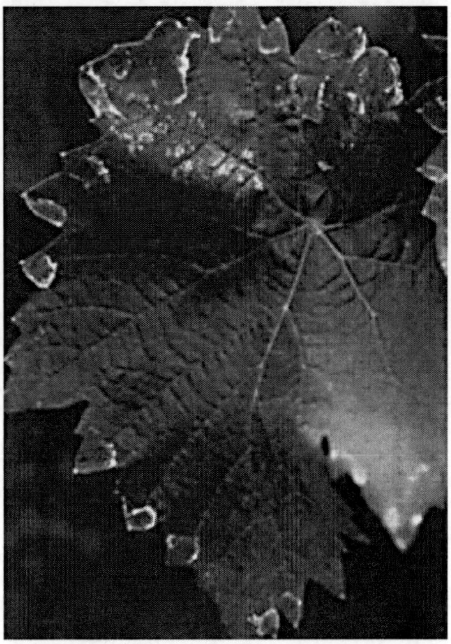

Figure 17. Excess of N.

Figure 18. Deficiency of P.

Figure 19. Deficiency of P.

Figure 20. Deficiency of K (the end of the summer).

Figure 21. Deficiency of K (the end of the summer).

Figure 22. Deficiency of Ca.

Figure 23. Excess of Ca.

Figure 24. Deficiency of Mg.

Figure 25. Deficiency of Mg.

6. Physiological Role of Microelements

Micronutrients are essentially as important as macronutrients to have better plant growth, yield and quality. Various physical and metabolic functions are governed by these mineral nutrients in the grapevine. They are required only in trace amounts, which are partly met from the soil or through chemical fertilizers or other sources (Hummes et al., 2019; Korchagin et al., 2020).

The most important factor controlling soil micronutrient availability for plant utilization is soil pH. In addition to the pH reaction of the soil to form an aqueous, the presence of carbonates ($CaCO_3$) in the soil, low soil aeration, soil compaction, low organic matter content, low temperature in the root zone, poor biological properties, and inadequate soil N status are also important factors controlling the supply of micronutrients for plant nutrition (Likar et al., 2015; Romic et al., 2012; Serrano et al., 2017). Trace elements such as Fe, Zn, and Mn are important activators of many essential enzymes involved in the metabolism of grapevines.

Microelements play a very important role in the biochemical processes that take place in plant cells, they participate in oxidation-reduction processes, they are part of enzymes, vitamins and hormones, and they influence the synthesis of carbohydrates and their migration from the leaves to the growing parts as well as to the fruiting organs. They also form part of complex organomineral compounds, which is one of the important factors for determining the direction and flow of physiological processes, as well as the dynamism of living matter as a whole. They have the role of biocatalysts. They actively participate in many complex processes in the plant: photosynthesis, respiration as well as protein synthesis. They influence the migration and redistribution of mineral elements in the plant.

Table 16. Microelements are required for healthy plant growth, their role in plant metabolism and symptoms associated with deficiencies (Reynolds & Braun, 2022)

Microelements	Pathway	Enzymes	Symptoms
Manganese (Mn)	Respiration	Some dehydrogenases, decarboxylases, kinases, oxidases and peroxidases	Reduced sugar and cellulose content, increased drought sensitivity, reduced fertility
Coopre (Cu)	Electron transport	Ascorbic acid oxidase, tyrosinase, monoamine oxidase, uricase, cytochrome oxidase, phenolase, laccase, and plastocyanin	Unlignified cell walls, permanent wilting and limp leaves
Boron (B)	Cell division, growth and membrane function	Synthesis of uracil, cell wall structure	Problems related to cell wall formation including reduced cluster and root growth, infertility
Zinc (Zn)	Electron transport and auxin biosynthesis	Alcohol dehydrogenase, glutamic dehydrogenase, and carbonic anhydrase	Interveinal chlorosis, and necrosis particularly in older leaves
Molybdenum (Mo)	Nitrogen use	Nitrogenase, nitrogen reductase	Nitrogen deficiency, chlorosis and necrosis on leaf margins. Leaves become pale and malformed.
Chlorine (Cl)	Photosynthetic reactions	/	Poor germination, chlorosis and necrotic lesions

The use of microelements has a positive effect on pollen germination, and then on fertilization, especially if they are applied foliarly before flowering. For this reason, when there is a lack of microelements in the nutrition of the grapevine, unfavourable conditions are created for the germination of pollen, fertilization and formation of the fruits, which on the other hand is the reason for the increased dropping of the flowers and the appearance of red clusters.

In their presence, plants absorb larger amounts of mineral substances. Microelements have a positive influence on the growth dynamics of the assimilation surface. They cause an exuberant growth of the clusters in length and thickness. They affect the increase of the leaf surface, especially the young leaves that develop quickly. In this way, the general habit of the grapevine is increased. The assimilation surface is one of the main factors for increasing the yield and obtaining a better quality of the grapes (Table 16).

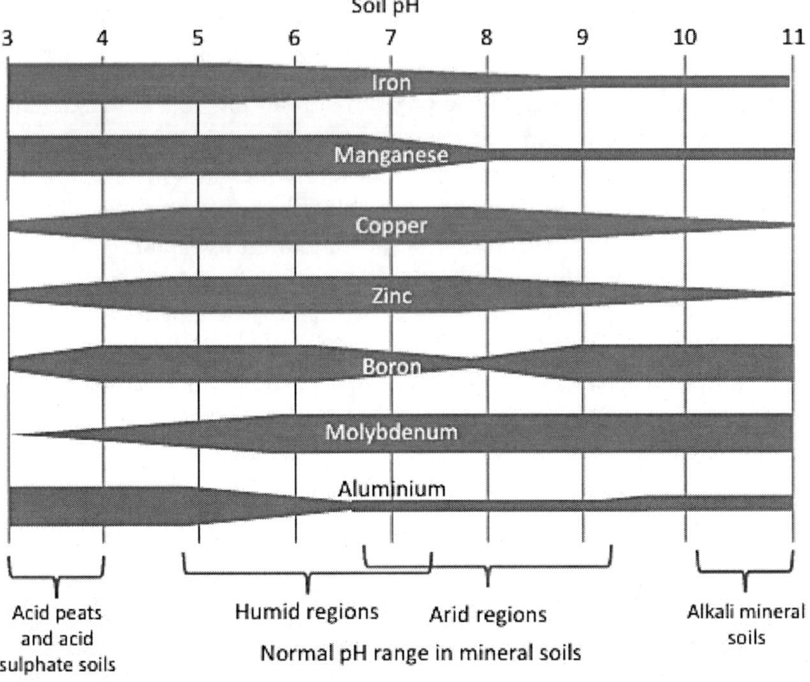

Figure 26. Diagrammatic representation of the relationship between soil pH and micronutrient availability (Reynolds & Braun, 2022).

They improve the maturation of clusters. It is known that after the end of vegetation, the grapevine goes into a state of winter dormancy. This condition is preceded by processes that were aimed at the maturation of the clusters, the formation of buds, the accumulation of reserve nutrients and the reduction of the amount of water. All these processes are greatly influenced by the microelements that influence the intensification of the accumulation of mineral substances, their migration and distribution in the individual organs of the grapevine, on which the degree of maturity of the clusters and other organs depends. They influence an increase in the amount of sugars in the grape berries, and later in the must.

Participating in the physiological processes, microelements affect the increase in the concentration of cell juice and the amount of dry matter in the grapevine, thus increasing the resistance of the grapevines as well as the resistance of the clusters to low winter temperatures.

Micronutrient deficiencies can lead to a wide range of alterations in normal plant growth and development. Visual symptoms are usually only apparent under extreme deficiency, but mild deficiencies can result in substantial reductions in grain yield. Given the variable role of these elements, the symptoms of deficiency also vary greatly (Reynolds & Braun, 2022).

Several factors can lead to micronutrient deficiency in plants including low levels of the nutrients in the soil and low mobility or availability of the nutrients due to low solubility in the form required for uptake. Soil-microbe interactions can also influence the availability of the micronutrients. Where free $CaCO_3$ is abundant in the soil chemistry, this can fix micronutrient cations, at a high soil pH the solubility of many micronutrients is reduced, and replenishment can be low if there is little organic matter in the soil. The impact of pH on nutrient availability is presented in Figure 26.

Applications of foliar nutrients, based on both NPK fertilizers and microelements, increased the incidence and severity of the expression of esca's foliar symptoms. In the case of applications of NPK fertilizers, the increase was associated with greater vegetative growth as a consequence of higher plant metabolic activity in the treated plants (Calzarano, 2023).

The optimized technology for the leaf treatment application of the chelated micronutrient fertilizers has a pronounced positive effect on the phytosanitary vineyards' condition and the grape's productivity. This occurs primarily due to the optimization of plant metabolism - maintaining a higher level of photosynthetic activity, leaves hydration during the season of critical growing and the proline accumulation in the leaves, which plays the role of a stress protector and ultimately affects the disease development and contributes to the formation of the complex indicator, such as productivity (Yurchenko & Artamonov, 2020).

The lack of microelements affects the development of the grapevine, especially the development of the berries. Their presence in the berries, and later in the must and wine conditions many significant physico-chemical processes. As a consequence of the lack of microelements in the nutrition of the grapevine, the structure of the buds can be deformed, and in some cases, more clusters can develop from one bud. These clusters are usually poorly developed.

Generally, microelements are applied directly to the vegetative part via foliar sprays, and these nutrients interact with each other to affect several pathways in the plant (Kumar et al., 2021). A successful fertilizer management program depends on the selection of the correct fertilizer to address the specific deficit observed in the vineyard, the correct rate of nutrient requirement and the timing for application of the fertilizer (Rahaman et al., 2019). Collectively, providing several benefits in terms of soil and environment protection, while enhancing the grape quality and yield.

In addition to the general characteristics, each microelement has its specific physiological-biochemical role, the knowledge of which is of great importance for the proper growth, development and fruiting of the grapevine.

6.1. Iron (Fe)

The grapevine absorbs iron from the soil solution in the form of Fe^{2+} and Fe^{3+} or as a chelate complex. In the soil solution, iron is usually present in an optimal amount, but its absorption is often difficult due to the high pH value, disturbed metabolism in the plant, as well as the influence of antagonistic elements.

Iron ions play a fundamental role in plants: they are constituents of the electron transport chains both in mitochondria and chloroplasts. However, the shortage of physiologically active iron leads first of all to a decreased number of thylakoid membranes per chloroplasts, unaffecting other iron-containing organelles (e.g., mitochondria and peroxisomes). The dysfunction of chloroplasts results in a rapid inhibition of chlorophyll formation, leading to leaf yellowing due to the decrease in the level of photosynthetic pigments (Rustioni et al., 2017).

The role of iron in the metabolic process has a special significance in the grapevine, and this is due to the ability of iron to change valency. It is part of the enzymes catalase, cytochrome oxidase, peroxidase, and nitrate reductase and in them, it has an important oxidoreduction role as a result of its ability to pass from ferrous to ferric cation. Although it is not included in the composition of chlorophyll, iron still participates in the photosynthesis process. It is a constitutive element of many enzymes involved in the transport of energy, in the formation of lignin, has an important role in nitrogen metabolism, and respiration, and is also a redox catalyst in the reduction of nitrates to nitrites and sulfates during the transport of electrons, etc. The physiological significance of iron is that in its absence it is not possible to synthesize chlorophyll, nor to initiate the activity of many important enzymes (catalase, peroxidase, cytochromes). Even that, it cannot regulate the necessary content of nitrogen and carbohydrates in the tissues of the grapevine.

6.1.1. Deficiency of Iron

Iron deficiency is most often observed on alkaline soils with $pH > 7$, as well as on soils rich in potassium and clay. In addition, deficiency may occur as a result of high phosphorus content in the soil, soil compaction, high groundwater levels, and increased bicarbonate and sodium content in the water used for irrigation. Only in exceptional cases can the cause of deficiency be the low content of iron in the soil itself.

The use of high doses of phosphorus fertilizers also affects the reduction of iron availability (insoluble Fe-phosphates are formed). The high content of calcium and phosphates at a high pH value of the soil contributes to the binding of iron with other ions, which transforms it into unavailable forms.

Furthermore, the physiological response of the plant to the iron deficiency (decrease in chlorophyll synthesis and difficulties in iron translocation) produced specific adaptations resulting in characteristic pigmentation distribution among leaves within the same cluster and veins and interveinal areas of the same leaf. From a physiological point of view, the compensation effect of basal leaves highlighted in iron-deficient plants appeared particularly interesting (Rustioni et al., 2017).

In the case of iron deficiency, iron chlorosis occurs in the grapevine, which is manifested by a change in the colour of the leaf mesophyll between the green nerves. It occurs as a result of an insufficient amount of iron in chloroplasts and loss of function. Chlorosis occurs first on the top leaves, first on the young leaves and then on the older ones. During the first stage of emergence, the nervature is green, and later it turns yellow. Young leaves first become light yellow, later turn yellow, and sometimes become completely white. The younger the leaves are, there are the greater iron deficiency symptoms. If the lack of iron persists, necrosis appears at the edges and leaves fall. The reason for this phenomenon is that iron is poorly mobile and does not reach the younger parts of the plant at the same time. Fe-chlorosis is similar to other chlorosis (Mg-chlorosis, Mn-chlorosis), and a certain difference can be determined if it is noticed at the beginning of the appearance. If symptoms of iron deficiency appear on the leaves, then the amount of phosphorus, copper, zinc, etc. increases in them. Consequently, when chlorotic symptoms in the vineyards develop, fruit yield and quality can be severely reduced in the current season as well as next season fruiting as fruit buds poorly develop. Foliar Fe application has been reported to effectively treat Fe-deficient syndromes and chlorosis in grapevines (Zebec et al., 2021).

Nevertheless, the deficiency of different mineral nutrients (e.g., magnesium, nitrogen) could affect the concentration of photosynthetic pigments, resulting in leaf yellowish. In these cases, the symptoms can be differentiated by the distribution among the clusters and within the leaves. Unlike nitrogen, which leads to a quite uniform leaf yellow colour, iron and magnesium develop pronounced interveinal chlorosis. These two deficiencies could be differentiated by symptom localization: iron deficiency occurs first on the youngest leaves (Rustioni et al., 2017).

Table 17. Symptoms of iron (Fe) deficiency and excess (Moyer et al., 2018; Stojanova, 2023)

	Deficiency	Excess
Symptoms	• Intervenal chlorosis (yellowing) of the leaves.	• The leaves are intensely green; • Antagonism with phosphorus and manganese.
Where (on the plant)	• The first symptoms appear on the young leaves.	• On the leaves.
Where (in the vineyard)	• Rarely on acidic soils; • On carbonate soils with a high pH value; • In vineyards with high groundwater, poor drainage or excessive irrigation.	/
When	• At the end of the summer.	/

The most typical symptoms of chlorosis occur in the absence of sufficient amounts of available forms of Fe and Mg. Symptoms of chlorosis are manifested by a change in the colour of the leaf mesophyll between the green nerves. The most common cause of chlorosis in plants is the occurrence of high amounts of calcium in the soil (carbonate soil). With a greater deficiency, yellowing also occurs at the tips of the green shoots (Table 17).

6.1.2. Excess of Iron

Iron excess rarely occurs in grapevines. If it occurs, the growth of all vegetative organs is inhibited, the leaves become dark blue-green and necrotic spots and necrosis gradually appear, especially noticeable on the edge of the leaf, and later the leaves become dry and fall. Due to antagonism, an excess of iron leads to a deficiency of phosphorus and manganese.

6.2. Manganese (Mn)

Plants absorb manganese in the form of Mn^{2+} ions and the form of Mn-chelate. The most accessible is the Mn^{2+} cation. Because of the antagonism with Mn^{2+}, its uptake can be affected by other divalent cations, especially Ca^{2+} and Mg^{2+}. Most soils are sufficiently supplied with manganese in the form of various oxides, hydroxides and salts. Manganese in the form of oxides is poorly mobile but is mobilized by bacteria in the soil. Due to its variable valency, the role of manganese is closely related to its participation in oxidation-reduction processes. Manganese deficiency can occur in coarse-textured alkaline soils where it can be leached out of the soils. This can be a serious issue in regions where wheat is grown in rotation with rice and inundation of the soil can leach Mn into deep soil layers (Reynolds & Braun, 2022).

It is mostly present in the young leaves, roots and other vital organs of the grapevine, while it is present in smaller amounts in the grapes. Manganese is not a constitutive element, but it has a significant participation in metabolic processes as an activator of a large number of enzymes that are involved in the processes of respiration, and synthesis of amino acids, lignin or hormones. Its role is similar to that of magnesium, which it can replace in a non-specific way by activating decarboxylase and dehydrogenase in the Krebs cycle. Manganese plays an important role in the synthesis of chlorophyll and nitrogen metabolism and is present in soil as exchangeable manganese or manganese oxide. Manganese is directly involved in the synthesis of fatty acids, in the neutralization of toxic oxygen radicals, and in the reduction of nitrates to ammonia (James et al., 2022).

The role of manganese is irreplaceable in the light phase of photosynthesis, in the transport of electrons and in the process of photooxidation (photolysis) of water (Ghorbani et al., 2019). It also enters into the composition of chloroplasts. It has the role of a biocatalyst, it stimulates growth, and the creation of chlorophyll A and B, and it especially affects the creation of carotenoids. It plays a role in the photolysis of water to hydrogen and oxygen. The physiological role of manganese in grapevine is consistent with the influence of magnesium. Manganese participates in metabolism, activates numerous enzymes in the processes of phosphorylation and helps the accumulation of carbohydrates and proteins. When the plant is well supplied with manganese, the need for N, P, K, and Ca is reduced without affecting the reduction of yields, which is why manganese is extremely important for the economic use of other elements in the soil.

Manganese has the role of a biocatalyst. It activates catalase and lipase and accelerates the action of hemosin and trypsin. It also catalyzes the process of biological fixation of nitrogen from the air.

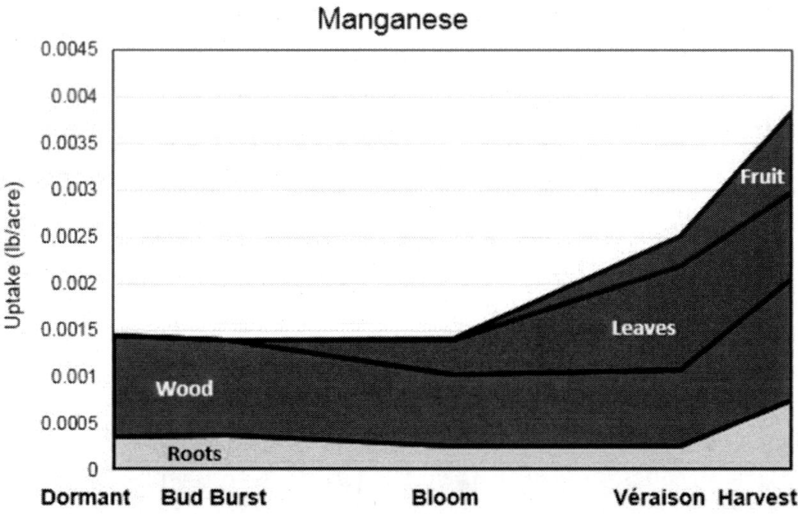

Figure 27. Timing and placement of manganese uptake by grapevines (Moyer et al., 2018).

It is a component of many enzymes, and it also enters into the construction of enzymes that catalyze oxidation and reduction reactions. It has the role of a growth stimulator, by stimulating the creation of chlorophyll A and B, and it has a special influence on the creation of higher oxidized carotenoids (xanthophylls), as well as carotene. It has a significant role in the photosynthesis process because it has been proven that the absence of manganese leads to the reduction of sucrose and reducing sugars in the plant, the migration of assimilates from the leaves to the root and other organs is accelerated, and it also can reduce the intensity of photosynthesis in the afternoon.

The role of manganese in plant tissues is variable depending on the nitrogen nutrition of the grapevine. Manganese functions as an oxidizing factor in the nutrition with ammonia nitrogen and as a reducing factor in the nutrition with nitrate nitrogen.

The negative impact of this element on the grapevine is related to the antagonism of manganese and iron. The ratio between manganese and iron is very important, which should be 1:2–2.5. If the ionic ratio of these two elements is in favour of manganese, the share of iron in the tissues decreases, regardless of whether Mn replaces Fe, or whether manganese stimulates the oxidation of Fe^{2+} – a physiologically more active form of Fe, which leads to disturbances in enzyme activity and the appearance of iron chlorosis symptoms.

Some antagonism also exists between manganese and calcium. As a result of calcium's ability to cause poor absorption of manganese, poor migration of this element to the leaves occurs (Figure 27).

The optimal presence of manganese conditions higher yield of grapes, faster ripening, reduction of acidity and creation of a larger amount of sugars, especially fructose.

6.2.1. Deficiency of Manganese

Manganese deficiency occurs in soils with an alkaline pH value where manganese passes into a trivalent form that is unavailable to the grapevine. Manganese deficiency can also be caused on sandy soils with a low content of organic matter, in drought conditions (increasing soil moisture increases the availability of Mn), on soils where calcification has been carried out with high doses of calcium, with high iron, copper and zinc content in the soil (James et al., 2022; Stojanova, 2023). Difficult uptake of manganese occurs in soils with a high content of Fe ions.

In the case of manganese deficiency, a physiological disease, manganese chlorosis, appears, which manifests itself on the edges of young leaves spreading towards the main nerve, so that the leaf has a mosaic appearance. The chlorotic part is clearly separated from the healthy part. The tissue around the nerve is green. The leaves first become light green, and later light brown, circular chlorotic spots are formed on them. At first, chlorotic spots appear only on the parts of the leaf that are distant from the main nerve. Later, due to the death of an increasing number of cells, the spots gradually spread and cover the entire leaf. The main nerve and the area around it remain undamaged for a long period. The symptoms of manganese chlorosis are very similar to the symptoms of magnesium and calcium deficiency. Grain growth and ripening are significantly reduced.

6.2.2. Excess of Manganese

Symptoms of excess manganese occur in conditions where manganese is present in higher soil content, as well as in acidic soil pH. The increased content of manganese in the nutrient solution always hurts the grapevine: it accelerates the destruction of the chloroplast ultrastructure, which causes a decrease in the intensity of photosynthesis and the accumulation of carbohydrates. At the same time, the leaves, leaf stalks and the nervature of the leaf are deformed. The negative impact of this element on the grapevine is related to the antagonism of manganese with iron, molybdenum and magnesium. Excess manganese blocks the uptake of iron, causing symptoms of iron chlorosis in grapevines. Visual symptoms are manifested as chlorotic and brown spots on old leaves. The order of occurrence of the symptoms of excess manganese is the opposite of the symptoms of iron deficiency chlorosis (Table 18).

6.3. Copper (Cu)

Plants take up copper as a Cu^{2+} cation or in the form of Cu-chelates. The adoption process is active. The physiological function of copper is due to its polyvalency.

The content of copper in the grapevine is low. It occurs as Cu^{2+} or as elemental Cu. It belongs to the group of heavy metals and can be displaced by many other cations. In the grapevine, it is mostly represented in the root, but it is also present in the chloroplasts.

Copper is a structural component of many enzymes: cytochrome oxidases; ascorbic acid oxidases (regulates the ascorbate/dehydroxyascorbate redox system); phenol oxidases; hydroxylases (transforms phenylalanine into alanine); oxygenases; galactooxidases (participates in the oxidation and breakdown processes of carbohydrates) (Miotto et al., 2014).

It affects the metabolism of nitrogen compounds so that with the increase in the amount of nitrogen, the amounts of proteins and the content of nucleic acids also increase.

It enables intensive breathing and affects the increase in the amount of carbohydrates in the plant. Its presence activates the redox potential and affects the migration of iron both in the soil and in the plant.

It favourably affects the stabilization of chlorophyll molecules, protecting them from destruction, and slowing down the aging of leaves. It affects the synthesis of nucleic acids, as well as the resistance of plants to low temperatures, droughts and diseases, participates in the processes of photosynthesis, especially in the light phase and in the turnover of substances through participation in the structure of enzymes, mostly in Cu-oxidase.

The effect of copper is particularly expressed on the leaves through the strengthening of chloroplasts. It affects the amount of chlorophyll in the leaves and its stability, so increasing the amount of copper in the leaves also increases the amount of chlorophyll in the leaves. All this affects the continuation of assimilation until the end of vegetation without a significant reduction in the intensity of photosynthesis.

Table 18. Symptoms of manganese (Mn) deficiency and excess (Moyer et al., 2018; Stojanova, 2023)

Symptoms	Deficiency	Excess
	• Intervenal chlorosis of older leaves. The tissue around the main and secondary nerves remains dark green; • A major deficiency results in interveinal necrosis of older leaves.	• Deformation of leaves, petioles and leaf veins.
Where (on the plant)	• On the basal leaves.	• Similar symptoms with copper deficiency.
Where (in the vineyard)	• Soils with a high content of calcium carbonate; • Sandy soils or areas with heavy rainfall.	• In soils with high manganese content and acidic soils.
When	• In the mid-season.	/

6.3.1. Deficiency of Copper

Deficiency of copper in the mineral nutrition of grapevines rarely occurs because pesticides used to prevent some grapevine diseases contain copper. The eventual lack of copper causes necrosis of the vegetative organs, wilting, bending and necrosis of the leaves, death of the top parts, and sometimes the whole plant. The first symptoms appear on the youngest leaves. Cluster growth is reduced. In the absence of copper in the grapevine nutrition, white spots appear on the leaves and there is a reduction in the growth of the plant. The main visual symptoms of excess Cu are the impaired growth of the roots and clusters, nutrient deficiency, chlorosis, and, in more severe cases, tissue necrosis and plant death (Miotto et al., 2014).

These symptoms are caused by the direct and indirect action of the Cu in plants. The reduction in the growth of the roots results in less exploration of the soil by the roots. The effect of Cu toxicity in the roots is reflected in indirect symptoms, such as the reduction of branch growth and the chlorosis caused by the generalized deficiency of nutrients and water (Yruela et al., 2000).

6.3.2. Excess of Copper

It is rare for there to be an excess of copper in the soil. The effects of excessive Cu in soils on the nutrition of grapevines are unclear. Young grapevine plants grown in soils newly contaminated with Cu demonstrate reduced growth in the roots and clusters, leaf chlorosis and Cu accumulation in the roots. However, a small amount of Cu is translocated to the branches (Toselli et al., 2009). In contrast, adult grapevines during the productive season do not exhibit visual symptoms of toxicity caused by Cu but can uptake and accumulate Cu in the perennial and annual organs (Lai et al., 2010). Grapevines remain productive for decades in soils with high Cu concentrations, and the plants are often subjected to spraying with Cu-based fungicides. The uptake of Cu from the soil and the Cu-containing fungicide sprayed directly on the aerial parts of the plant represent two sources of this metal for the plant. Under conditions of high levels of Cu in the soil and the spraying of Cu-based fungicides, grapevines can exhibit increased uptake and accumulation of Cu and toxicity symptoms of excessive Cu in the tissues. In addition, Cu accumulation in the roots may affect the uptake of other nutrients, causing nutritional problems for the plants (Miotto et al., 2014).

In excess of copper, older grapevines have a well-developed root system that can accumulate and inactivate large amounts of copper. The first symptoms of excess copper appear on old leaves. Chlorosis spreads from the tips and edges to the middle of the leaf and leaf petiole, followed by leaf death. In young plants, excess copper can cause damage in the form of growth arrest and chlorosis (Table 19).

6.4. Boron (B)

Plants take up boron from the soil by passive diffusion. In conditions of optimal concentration or with an excess of boron in the soil, uptake takes place passively and boron is easily transferred through the plasmalemma. In the case of reduced boron concentration in the soil, the process is still active and uptake takes place through specialized boron transporters, consuming energy from ATP. Further transport from the cells of the endodermis to the elements of the xylem takes place similarly.

This element is of great importance in the nutrition of the grapevine. Boron has a great influence on the growth, flowering, fertilization and yield of grapes. Also, B is essential for the development of plant meristem and the metabolism of amino acids, proteins, carbohydrates, Ca, and water. It contributes to better utilization of Ca in plants (Stefanello et al., 2021). Boron fertilizers increase the concentration of boron in the plant during the entire vegetative period, which favourably affects the development of the grapevine, increasing yields and obtaining better quality of the grapes. Whereas, B accumulation in plants increases with increasing soil K (Bredun et al., 2021). After the foliar application, boron is quickly absorbed by the leaf and translocated to the generative organs of the grapevine (Table 6). Its most important role is in the metabolic complexes that are important in nutrition and in increasing the fertility and germination of pollen grains and the development of the pollen tube, thereby ensuring good conditions for normal fertilization. Boron is especially important for maintaining tissue hydration. It affects the regulation of respiration energy, because with boron deficiency, plants breathe intensively, both above-ground organs and roots.

The ability of boron to form complex organo-boron compounds with D-fructose, D-galactose, D-glucose, D-arabinose, glycerin, pyridoxine and with some flavones indicates that it has a positive effect on carbohydrate metabolism, primarily in the synthesis and migration of sucrose, as well as in the synthesis of proteins, which is important for grapevine as a crop that produces a large number of carbohydrates.

Table 19. Symptoms of copper (Cu) deficiency and excess (Moyer et al., 2018; Stojanova, 2023)

Symptoms	Deficiency	Excess
	• Small leaves, short internodes.	• Chlorosis occurs, which spreads from the tips and edges to the middle of the leaf and the petiole, and then leaves die. • In young plants, excess copper can cause stunting damage.
Where (on the plant)	• Symptoms occur on the youngest leaves.	• On the old leaves.
Where (in the vineyard)	• Shallow, sandy and high pH soils.	• During long-term use of preparations for protection of the grapevine, a high concentration of Cu in the soil occurs.
When	• At the end of summer.	

The influence of boron is at the cellular level: it contributes to the construction of the structure and stability of the cell wall and accelerates the differentiation of the cambial cells towards the xylem and phloem. In the absence of boron, the transport of the products of photosynthesis slows down, which leads to their accumulation in the leaf and the slowing down of new syntheses (Figure 28).

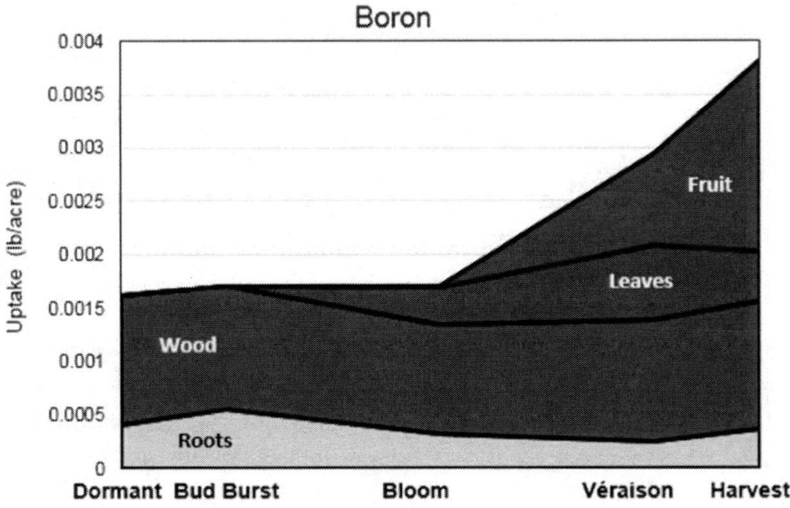

Figure 28. Timing and placement of boron uptake by grapevines (Moyer et al., 2018).

In the presence of a sufficient amount of boron, plants use other nutrients better and more economically, primarily nitrogen, phosphorus and potassium.

6.4.1. Deficiency of Boron

The lack of boron is primarily due to its high mobility in the soil and to the rapid migration from the upper to the lower layers of the soil, i.e., below the level of the root system. The appearance of boron deficiency is generally observed on alkaline and sandy soils, poor in organic matter, as well as on heavy and cold soils with a very low pH value (4.5–5.0), but also strongly carbonated soils (pH > 8). On alkaline soils, in conditions of excess potassium and calcium, there is a blockage of boron and problems in its absorption. This is especially pronounced in conditions of dry years. In conditions of poor supply of soil with boron, its deficiency in plants is similar to deficiency due to drought, high soil pH or excessive calcification.

Boron deficiency shows various symptoms including stunted growth, such that grapevine internodes are shortened displaying a zig-zag pattern, meristem distortion, death of cluster tips, interveinal chlorosis of older leaves and poor flowering (Peacock & Christensen, 2005).

With a lack of boron, there is a reduction in the amount of chlorophyll in the leaves. This decrease in chlorophyll correlates with the age of the leaves, so the younger the plant, the greater the decrease in chlorophyll. The decrease in the amount of chlorophyll has no significant impact on the intensity of photosynthesis, but conditions an intense accumulation of soluble carbohydrates, both in the leaf and in the root. The lack of boron leads to a decrease in the intensity of protein biosynthesis, and the accumulation of amino acids increases.

With a lack of boron, changes occur in the anatomical structure of the plants, which appear as a result of the unusual division of the cambium, which forms unequal cells. Some cells are very large, elongated and with large nuclei, while others are too small. Degenerative decay of the cells of the cambium is also observed, during which a tissue with a dark brown colour appears. After the degeneration of the cambial layer, changes occur in the basic parenchyma, especially in the part bordering the xylem. In tracheids, when there is a lack of boron, the internal space is reduced, which is clogged with resinous matter. The anatomical elements of the root show similar degenerative changes. With a lack of boron, the root system is poorly developed, and if the deficiency is present for a long period, it can lead to complete stunting of the root.

The appearance of non-functional flowers is also possible. Pollen germination and pollen tube growth are hampered without the presence of an optimal amount of boron. There are numerous examples when, despite good flowering, fertilization does not occur, which is associated with the lack of boron in the grapevine's nutrition.

Boron is immobile in the plant, so deficiency symptoms usually appear at the site of growth. It should be emphasized that the grapevine reacts very quickly to the lack of boron in the nutrient solution, due to which, in conditions of its weak or complete absence, the growth of all organs is significantly reduced, the normal development of the stem, root and branches is shortened.

Boron deficiency symptoms are manifested by the appearance of round necrotic spots with a yellow colour on the leaves of white varieties, and a red colour on the leaves of red varieties. Circular necrosis first appears on the tips of the leaves and later on in the middle of the leaves. The leaves are dried starting from the edges and towards the median nerve.

With a lack of boron, the growth of green shoots stops very quickly, the internodes are shorter, and the ends of the leaves grow bushy. Flowering and fertilization are difficult because the pollen's viability is weak, so the bunches remain red. Both the leaf and the kernels show a yellowish necrosis, and the kernels are often shrivelled and remain small. When they reach a size of 2 to 4 mm, they become darkened, which resembles the appearance of downy mildew or powdery mildew. With an acute lack of boron in the grapevine, the formation of seeds in the berries is absent.

Due to the lack of boron, the cambium and phloem decay, while the xylem remains undamaged. Due to the damage to the phloem, carbohydrates accumulate in the leaves, which causes disruption of protein synthesis, and this is the reason for an increased amount of soluble nitrogen.

Figure 29. Progression of B toxicity in the grapevine variety Chardonnay from left to right (Rogiers et al., 2021).

Table 20. Symptoms of boron (B) deficiency and excess (Moyer et al., 2018; Stojanova, 2023)

Symptoms	Deficiency	Excess
	• Reduced yield, small berries; • Necrosis of leaves and clusters; • If the previous autumn and winter were dry, the basal leaves have a wrinkled appearance with pronounced nervature.	• Necrosis on the edges of the leaves.
Where (on the plant)	• Leaf symptoms occur near the tip of the cluster (boron deficiency in the current season); • On the basal leaves (if the previous autumn and winter were dry).	• The first symptoms appear on older leaves.
Where (in the vineyard)	• Alkaline, acidic, sandy soils, in areas with a lot of precipitation, soils with a lack of organic matter; • Dry conditions.	
When	At the end of summer.	At the end of summer.

One of the typical symptoms of boron deficiency is the appearance of a strong thickening of the internodes and the formation of long tendrils. This phenomenon is observed in summer (Table 20).

Excess of boron – Excess boron in the soil or the irrigation water has a toxic effect on the grapevines. There is antagonism between boron and calcium in the soil. An excess of boron prevents growth and leads to the death of the leaves (Figure 29).

6.5. Zinc (Zn)

Zinc is mostly absorbed as a divalent cation Zn^{2+}, and in conditions of a higher pH value, it can also appear as a monovalent cation $Zn(OH)^{+}$. It can also be taken in the form of Zn-chelates. The absorption takes place actively, and at the same time, it has an inhibitory effect on iron, copper and manganese ions. It is mostly found in chloroplasts, cell walls, young leaves and roots.

This microelement is found in all parts of the grapevine. In buds, leaves and green shoots, zinc is significantly more abundant than one-year-old clusters, which are the least supplied with zinc. The hard parts of the bunch, especially the stalks, are richer in zinc.

After adsorbing, zinc is actively involved in the activity of enzymes where it has three functions, namely: catalytic function (as in carbonic anhydrases), co-catalytic function (two enzymes are bound) and structural function (when zinc is connected to four amino acid residues). Hence, zinc can influence many biochemical reactions. The role of zinc in the biosynthesis of DNA and RNA (RNA-polymerase) is of great importance. Also, through site-specific modifications, regulation of the chromatin structure, RNA metabolism and protein–protein interactions, the synthesis of proteins in the metabolism of auxins, affects the growth of the plant (it affects the synthesis of tryptophan), the stabilization of biomembranes, etc. (Daccak et al., 2022).

It is involved in chlorophyll formation and is an essential component of certain plant enzymes, e.g., in auxin formation. Presently, Zn is recognized as a vital component in several dehydrogenases, proteinases, Zn-containing enzymes, Zn-activated enzymes, protein synthesis, carbohydrate metabolism, tryptophan and indoleacetic acid synthesis, membrane integrity and lipid peroxidation (Navarro et al., 2021).

It participates in the construction of the enzymes carbonic anhydrases and dehydrogenases and affects the activity of ribulose-1,5-diphosphate, and carboxylases, in the uptake and transport of phosphorus, the activity of phosphatases, it affects the increase of resistance to diseases (through the influence of photosynthesis), resistance to drought (reduces transpiration) and resistance to low temperatures. It participates in the activity of phosphatases.

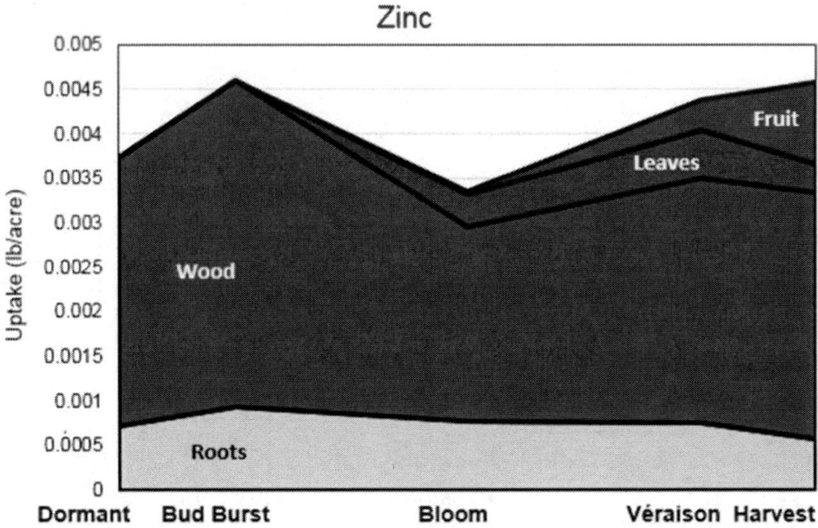

Figure 30. Timing and placement of zinc uptake by grapevines (Moyer et al., 2018).

Table 21. Symptoms of zinc (Zn) deficiency and excess (Moyer et al., 2018; Stojanova, 2023)

Symptoms	Deficiency	Excess
	• Reduced yield; • The berries remain small and do not ripen; • The leaves are small and pale in colour; • Short internodes.	• The leaves are small with reddish spots, and necrosis occurs at the edges.
Where (on the plant)	• The first symptoms appear on the old lower leaves.	• Both on old and young leaves.
Where (in the vineyard)	• Shallow, sandy soils, as well as soils with a high pH value, often deficient in Fe.	• On surfaces that are fertilized with low-quality organic fertilizer.
When	• In late spring and early summer.	/

The absorption of zinc and its activity are closely related to the amounts of phosphorus. The high amount of phosphorus in plants acts antagonistically to zinc, whereby zinc in the leaf becomes difficult to move, it is located around the nerves, and not on the entire leaf surface (Figure 30). Zinc is an essential component of various enzymes involved in energy production. It has an important role in the physiological processes of respiration, synthesis of carbohydrates, biosynthesis of chlorophyll, creation of phosphatides and vitamin C.

In the presence of zinc in grapevine, resistance to heat and drought increases (zinc reduces transpiration). It has also been established that it affects the acceleration of the growth of the root system and the retention of carbohydrates in it in the autumn, which explains the resistance to low winter temperatures. Zinc manifests a similar effect in terms of the resistance of the stems and clusters to low winter temperatures. This acquired resistance is conditioned by the influence of zinc on the colloidal-chemical properties of the protoplasm. Through its influence on photosynthesis, zinc increases the resistance of the grapevine to diseases.

6.5.1. Deficiency of Zinc

The lack of zinc in the mineral nutrition of the grapevine is rarely caused by a direct lack of this element in the soil but is caused by the influence of other factors, primarily by the increased content of phosphorus in the soil. The lack of zinc occurs more often on sandy, alkaline soils, on soils poor in organic matter and on soils that contain too high content of phosphorus, on soils with a pH value of 6.5 to 8.0, and the lack of zinc on carbonate soils is often associated with iron deficiency (Navarro et al., 2021). On carbonate soils, the amount of soluble zinc Zn^{2+} available to plants is very low. Zinc deficiency may result in stunted growth and development of small, undersized leaves with mottling between veins, clawed margins, and widened petiolar sinus (Ali et al., 2021).

Zinc deficiency usually occurs at the end of spring and at the beginning of summer in conditions of low temperatures and high humidity in the soil and air. The first symptoms of zinc deficiency are observed on the old lower leaves. In these leaves, there is a disturbance in the very uneven cells. They are very large, with large vacuoles and rough membranes, or they can be small, delicate and plastic. In severe zinc deficiency, the internodes remain short and the leaves are small and rough. The bunches that are formed have very sparse berries, are small and about the size of peas. With a lack of zinc, a greater amount of inorganic phosphorus accumulates in the clusters, a lack of starch occurs in the leaves, and at the same time, greater amounts of phenolic substances and phytosterols accumulate. In addition, the amount of auxin is significantly reduced, which causes the arrest of growth and development.

Among the many symptoms of zinc deficiency, the most common is the appearance of chlorosis (similar to the appearance of iron deficiency) in the area between the nerves of the leaf. The symptoms appear on the top parts of the clusters where the internodes are shortened. The leaves remain small and have a rosette appearance. The leaves are deformed due to the appearance of a large number of small yellow spots with a mosaic appearance, and parts of the leaf may dry out. It is manifested by the appearance of chlorosis between the nerves of the leaves. The surface of the leaf acquires a light green colour, and the symptoms of chlorosis and necrosis in old leaves are often secondary effects caused by the toxic effect of boron. The leaf is smooth and thick, and the petiole is shortened. The grapevine grows like a bush with weakly lush clusters and many lateral shoots. These symptoms are often equated with iron and manganese deficiency symptoms. When there is a lack of zinc in plant tissues, there is an accumulation of polyphenols, phytosterols or lecithin. This phenomenon occurs as a result of incomplete oxidation of proteins in the absence of zinc. The lack of zinc negatively affects the fertilization and binding of grains. The clusters are poorly developed. Very often low zinc is followed by low magnesium.

6.5.2. Excess of Zinc

Zinc excess rarely occurs and only on acidic soils. High amounts of phosphorus in the soil slow down the uptake of zinc. Also, zinc excess causes undesirable physiological, anatomical and morphological changes. Symptoms of zinc excess are necrosis of the marginal parts of the leaves and the appearance of brownish spots. This makes it difficult to absorb phosphorus and iron, which is why symptoms of deficiency of these two elements can also appear (Table 21).

6.6. Cobalt (Co)

This element is found in soil in small amounts. The grapevine adopts it as a divalent cation. The needs of the grapevine are small, but often cannot be met due to antagonism with iron, copper, zinc and manganese that hinder the uptake of cobalt. Its action in the plant is related to polyvalency. It usually participates in oxidation processes, increases the turnover of peroxidase, and enters the composition of vitamin B_{12}. It is mostly present in the top leaves.

6.6.1. Deficiency of Cobalt

In the case of cobalt deficiency, chlorosis occurs on the young leaves, but this is not a typical symptom (there are similar symptoms in nitrogen and iron deficiency).

6.6.2. Excess of Cobalt

No excess cobalt has been observed in natural conditions. The higher amount of cobalt reduces the absorption of iron, which is the reason for the appearance of leaf chlorosis. If an excess of cobalt occurs, the symptoms are similar to an excess of magnesium (Table 22).

6.7. Molybdenum (Mo)

Molybdenum is absorbed in the form of the molybdenum anion, MoO_4^{2-}. The function of molybdenum in biological systems is related to its ability to change valency (Mo^{4+}/Mo^{5+}). This allows molybdenum to participate in the transport of electrons, and in oxidation-reduction processes, it is part of many enzymes.

Table 22. Symptoms of cobalt (Co) deficiency and excess (Moyer et al., 2018; Stojanova, 2023)

Symptoms	Deficiency	Excess
	• Chlorosis.	• Similar symptoms with excess magnesium.
Where (on the plant)	• On young leaves.	• On the leaves.
Where (in the vineyard)	• In soils where Co antagonism occurs with Fe, Cu, Zn and Mn.	/
When	/	/

It is represented in small amounts in the organs of the grapevine, and mostly in the roots. The grapevine absorbs molybdenum in the form of molybdate through the root and the leaves. The presence of SO_4^{2-} ions hinders the absorption of this element.

As a metal complex, it participates in the structure of many enzymes: nitrate reductase, aldehyde oxidase, xanthine dehydrogenase and sulfide oxidase. These enzymes have great metabolic importance, especially in nitrogen metabolism. The most important is nitrate reductase, which has a significant role in nitrogen metabolism.

Table 23. Symptoms of molybdenum (Mo) deficiency and excess (Moyer et al., 2018; Stojanova, 2023)

Symptoms	Deficiency	Excess
	• Molybdenum deficiency has similar symptoms to N, Ca and Mg deficiency.	• Similar symptoms with Mo and B deficiency.
Where (on the plant)	• Sometimes only on young leaves, and sometimes on the whole plant.	• The grapevine grows poorly, clusters and leaves are thickened, yellowing young leaves. Chlorosis appears on the young leaves, which later spreads to other parts of the grapevine, changing from yellow to brown.
Where (in the vineyard)	• On acidic podzolic soils.	• An excess is rarely observed.
When	• It is rarely noticed.	/

The presence of molybdenum affects the accumulation of nitrogen in leaves and tissues, and an increase in osmotic pressure occurs. All this has an impact on the development of the root system, on the vegetative development of the grapevine, as well as on the increase in yields.

Molybdenum is of great importance for the growth process as well as for other processes because the enzyme aldehydroxydase regulates the synthesis of the plant hormone auxin. A series of physiological processes in the plant depends on this hormone, such as the growth of plant organs and resistance of plants to abiotic and biotic stress factors. In the presence of molybdenum, there is an increase in the amount of nitrogen in the leaves and the tissues, and an increase in osmotic pressure also occurs.

Another physiological role of Mo is its participation in the metabolism of phosphorus and iron, then in the synthesis of vitamin C and carbohydrates. A correlation has been observed between vitamin C and molybdenum in plants.

6.7.1. Deficiency of Molybdenum

Molybdenum deficiency is rare in grapevine nutrition. Symptoms rarely occur on young leaves and are not typical due to their great similarity with symptoms of other elements (deficiency of nitrogen, calcium and magnesium), as well as excess of manganese. The consequence of the deficiency of molybdenum is the appearance of chlorosis of the clusters and the appearance of necrosis of the leaves, delayed flowering and the appearance of longer internodes than normal. If molybdenum deficiency occurs, the intensity of nitrate reduction slows down, and this affects the deterioration of protein metabolism (Table 23).

6.7.2. Excess of Molybdenum

An excess of molybdenum rarely occurs, and if it occurs, it manifests itself with symptoms that are similar to the symptoms of a lack of molybdenum and boron. A high amount of molybdenum has a toxic effect on plants. Symptoms of excess molybdenum are reduced grapevine growth, while green shoots and leaves become thickened. Chlorosis occurs on young leaves, which later spreads to other parts of the grapevine, changing from yellow to brown.

The symptoms of deficiency and excess of individual microelements are illustrated in Appendix B (Figure 31 to 42).

Appendix B: Symptoms of Deficiency and Excess of Microelements (Stojanova, 2023)

Figure 31. Deficiency of Fe.

Figure 32. Deficiency of Fe.

Figure 33. Deficiency of Mn.

Figure 34. Deficiency of Mn.

Figure 35. Deficiency of Cu.

Figure 36. Excess of Cu.

Figure 37. Deficiency of B.

Figure 38. Excess of B.

Figure 39. Deficiency of Zn.

Figure 40. Deficiency of Mg and Zn.

Figure 41. Excess of Cl^-.

Figure 42. Excess of Cl^-.

Chapter 2

Influence of Macro and Microelements on the Quality and Biological Potential of Grapes and Wine

1. Basic Explanation

With the use of macro and microelements in the nutrition of the grapevines, in addition to increasing yields, favourable conditions are also created for obtaining better quality grapes. In wine-growing areas, especially where there are no favourable climatic conditions, by applying different methods of feeding the grapevines, it is possible to successfully influence the improvement of the chemical composition of the grapes.

The overall productivity of the grapevine is conditioned by the intensity of photosynthesis and the interrelationships of the metabolic processes. The yield of grapes depends on the photosynthetic activity (leaves) and the direction of the synthesized products more towards the bunches and less towards other organs. The relationship between grapevine growth and fruiting, as well as the relationship between grape yield and quality, depends on the compatibility of biotic and abiotic factors.

The quality of the grapes, the colour of the berries, the taste, the size, the firmness, etc. are varietal characteristics. Climatic, ecological and other environmental factors, planting density, cultivation form, pruning, etc., influence the listed properties. Agrotechnical measures are also of great importance, including nutrition, that is, fertilizing. The quality of the grapes is closely dependent on the ratio of fertility and vegetation. The lower this ratio, the better the quality of the grapes.

Depending on the applied doses and the ratio of the individual nutrients in the fertilizers, they can improve, or in the case of their unprofessional application, worsen the quality of the grapes and, later, of their processing. Of the macroelements, nitrogen, phosphorus, potassium, calcium and magnesium have the greatest influence on the quality of the grapes, and of the microelements, boron, manganese and zinc have a special influence on the quality of the grapes. Participating in physiological processes, microelements affect the migration of carbohydrates, increasing the concentration of sugars in grains. Their deficiency or excess can have a very unfavourable effect on the quality of grapes.

Due to the high fertility of the soil, excessive rainfall or excessive fertilization can lead to the extension of the vegetation, which on the other hand harms the quality of the grapes. In this case, part of the assimilates that are needed for the normal development and ripening of the grapes will be used for additional growth of the clusters and tendrils, which has a great impact on slowing down the ripening of the grapes and affects the reduction of the concentration of sugars.

If the soil is poorly fertile and if there is a drought, the vegetation of the clusters will end earlier, which also harms the quality of the grapes, because the grapevine is not able to provide a sufficient amount of nutrients for the normal development and ripening of the grapes. In this case, fertilization can influence the continuation of the vegetation, which creates conditions for the accumulation of optimal amounts of assimilates by the grapevines necessary for the proper development of the grapes.

For this reason, the application of optimal doses of fertilizers is very important, whereby in the phenophase the grapevine will not have an intense growth, but will be normally developed and will provide enough assimilatives for the ripening of the grapes, which has a positive effect on the quality of the grapes.

Correct and well-balanced nutrition has a favourable effect on the quality of grapes, while unilateral and untimely nutrition, especially the incorrect use of nitrogen and potassium fertilizers, can cause undesirable chemical, physical and organoleptic changes, which deteriorates the quality of grapes (Stojanova, 2023).

2. Chemical Composition of Grapes

Whether used fresh as a fruit or in the form of juice, wine, raisins and jam, grapes are one of the most consumed fruits worldwide. It is estimated that these quantities exceed 60 million tons annually worldwide, while the European part contributes 23 million tons. Grapes are rich in vitamins and minerals, carbohydrates, lipids, proteins and polyphenolic compounds, which are distributed between skin, seeds and pulp. Which metabolites will be present in grapes and in what quantities depends mostly on the grape variety, agroclimatic factors, grape maturity at the time of picking, soil composition and exposure to pathogens. In addition to the mentioned factors, the chemical composition of juice and wine is additionally affected by technological processing and procedures that, through a series of reactions (oxidation, polymerization, copigmentation), contribute to the overall complexity of the final product (Ali et al., 2010; Teixeira et al., 2013; Nowshehri et al., 2015; Hornedo-Ortega et al., 2020).

Achieving a good quality in grapes is an essential goal wherever it is grown; one of the important components that make the quality is the phytochemical content of the berries. Grapes contain many phytochemicals beneficial for human health, as well as amino acids, proteins, vitamins, and minerals. So, berries are efficiently used to increase the nutritional and energy value of the human diet (Tangolar et al., 2023).

The chemical composition of grapes, must and wine has been studied in detail since the 60s of the 20th century. With the development of analytical techniques, it is possible to discover many different compounds that have an impact on the overall quality of the wine, as well as on the characteristics of each wine. Progress in discovering the composition of grapes, must and wine and defining the role of various compounds in the formation of wine properties is accompanied by changes in viticultural practices that best correspond to new knowledge about the role of various substances in the formation of wine quality.

Potential bioactivity and medicinal properties have been attributed to all plant parts, especially pomace, clusters, stems and leaves, which are used in the formulation of dietary antioxidant supplements (Dani et al., 2010). In addition, grapevine leaves have a pleasant taste and can be used as fresh food, boiled or baked. Also, it is often found as a food supplement on the market. The chemical composition is greatly influenced by the variety, degree of maturation, climatic conditions and the location where the plants are grown (Taware et al., 2010; Teixeira et al., 2013).

Changes in the phenolic profile and antioxidant activity of the leaves of *Vitis vinifera* L. occur due to the influence of variety, time of picking (Katalinić et al., 2013), stress, drought (Krol et al., 2014) and powdery mildew infection (Taware et al., 2010). Various factors have been recognized that cause variations in elemental composition, such as soil type, fertilizer use, climate, irrigation practices, plant maturity stage, and different cultivars of the same species (Likar et al., 2015). Since ancient times, *Vitis vinifera* leaves have been used in medicine for various biological activities, including hepatoprotective, spasmolytic, hypoglycemic and vasorelaxant effects, as well as antibacterial, antifungal, anti-inflammatory, antinociceptive, antiviral and especially antioxidant properties. The juice of the leaves is also recommended as an antiseptic eye wash.

A large number of different compounds have been isolated from the fruits and leaves of the grapevine, which can be classified according to their chemical structure into several large groups. These are phenolic compounds, organic acids, vitamins, enzymes, carbohydrates, organic compounds with nitrogen, terpenoids and volatile compounds, waxes, lipids, polysaccharides and gums.

Chemical tests of the leaves showed the presence of several organic acids, phenolic acids, flavonols, tannins, procyanidins, anthocyanins, lipids, enzymes, vitamins, carotenoids, terpenes and reducing or non-reducing sugars. These discoveries have led to considerable interest in the leaves of *V. vinifera*, as a promising source of compounds with nutritional properties and biological potential. It also solves the disposal problems that arise due to the large amount of waste produced by the wine and juice industry (Monagas et al., 2006).

The chemical composition of grapes is very complex and variable and is one of the basic factors for determining the quality of grapes for fresh consumption and as raw material for processing in the production process of various alcoholic and non-alcoholic beverages. The chemical composition of grapes is not constant and can vary significantly depending on the variety, environmental conditions, cultivation measures, grape care and many other factors (Figure 43).

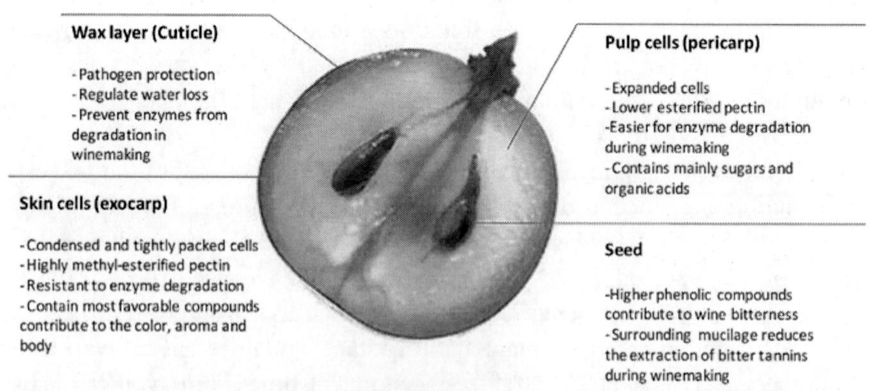

Figure 43. The biological anatomy and biochemical composition of a typical wine grape berry concerning extractable components (Gao et al., 2019).

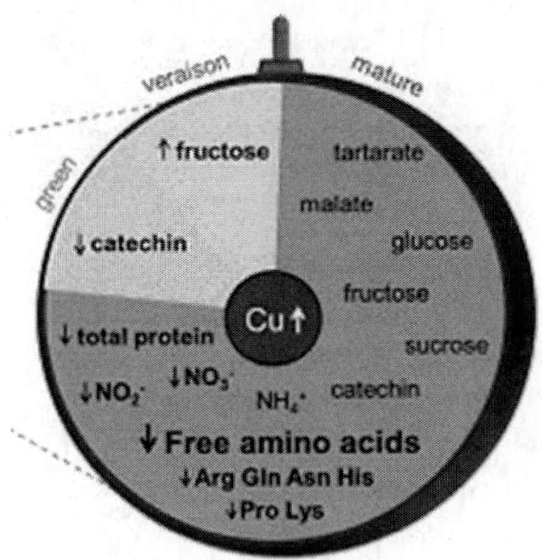

Figure 44. Chemical composition of grape barriers (Martins et al., 2014).

In some varieties of grapes, the variations of the chemical composition from year to year are variable or change depending on the conditions of production. Of all the substances that influence the chemical composition of grapes for processing into wine, the most significant are carbohydrates and total acids.

The quality of the grapevine depends mainly on its metabolites. Accumulation of metabolites is particularly sensitive to external conditions. The chemical diversity of grapevine varieties mostly refers to secondary metabolites, which belong to different phytochemical groups such as alkaloids, terpenes, antibiotics, volatile oils, sterols, saponins and phenolic compounds (Figure 44). Therefore, the grapevine is of great importance for the food, pharmaceutical, agrochemical and cosmetic industries (Ali et al., 2010).

Grapevine leaves are rich in minerals, vitamins, carotenoids and phenolic compounds. Flavonols and phenolic acids are the main classes of phenolic compounds in grapevine leaves (Dani et al., 2010), and flavan-3-ols and stilbenes are present in lower amounts. The synthesis of phenolic compounds is influenced by the developmental stage and external factors (Sun et al., 2006). In particular, ultraviolet radiation and infection with certain fungi can induce the synthesis of phytoalexins, which have been shown to play significant roles in human health. Phytoalexins have antioxidant properties, cardioprotective and anti-ageing activities and antimicrobial activity. Some authors reported that these compounds are synthesized only under stressful conditions and that in healthy leaves they are absent or present only at a trace level (Sun et al., 2006).

More and more attention is paid to the quantities, but also to the structure of the phenolic and aromatic compounds in the grapes that will be processed into wine.

2.1. Carbohydrates in Grapes

Carbohydrates are the most abundant ingredients in grapes after water. The total content of carbohydrates present in grapes is variable and depends on the biological characteristics of the variety, the growing conditions of the grapevines, the degree of ripening (warm weather conditions, good exposure to the sun and good ripening of the grapes help their formation, while low temperatures, poor exposure and incomplete ripening result in a lower amount of sugars), applied agrotechnical and ampelotechnical measures and from the application of optimal nutrition. Carbohydrate content ranges widely from 9 to 20% (Kennedy et al., 2000).

Most of the sugars are formed in the leaves with the process of photosynthesis, and a part is formed in the berries while they are green and contain chlorophyll. Green berries with their photosynthetic activity can satisfy up to 20% of their own needs, but since sugars are consumed in the process of respiration, their amount in green berries is small.

Almost the entire amount of sugars in grapes is made up of monosaccharides, i.e., the hexoses glucose (grape sugar) and fructose (fruit sugar). Sucrose is represented in a smaller amount. The must from grapes in a state of physiological maturity contains approximately equal amounts of these two sugars, that is, it is said that the so-called G/F index (ratio of glucose and fructose) is approximately equal to 1. The amount and mutual ratio of glucose and fructose are variable during the ripening of the bunches. At the beginning of grape ripening, the proportion of glucose is greater compared to fructose. Unripe grapes contain a larger amount of glucose, and overripe grapes contain more fructose, that is, as the grapes ripen, the ratio of these two carbohydrates decreases so that in ripe grapes their ratio is approximately equalized (Kennedy et al., 2000; Stojanova, 2023).

Knowing the ratio of glucose and fructose in the grapes being processed is significant, especially in the processing of overripe grapes in the production of natural sweet wines.

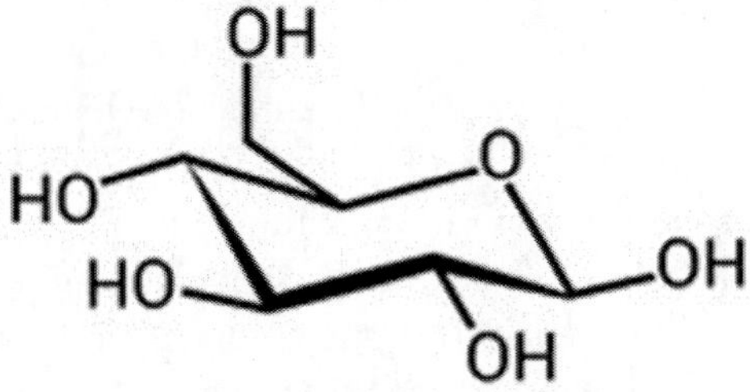

Figure 45. Beta-D-glucose in grapes.

Grapes infected with *Botrytis cinerea* (grey mould) contain a higher amount of fructose concerning the amount of glucose (Figure 45). Compared to glucose, fructose is a much more suitable medium for the development of bacteria and yeasts (the cause of various wine spoilage). Grapes always contain certain smaller amounts of sucrose (usually 3–5, rarely up to 15 g/L) and a small amount of pentose (about 1 g/L).

Sucrose (Figure 46) as a disaccharide is not subject to alcoholic fermentation, but under the influence of acidic environmental conditions and yeast enzymes, it is quickly broken down into its constituent parts - glucose and fructose, which yeasts use during fermentation. Thanks to this, sucrose can also be used to increase the sweetness of wine, which in years with bad climatic conditions is not at the level required for wine production. The pentoses that grapes usually contain in concentrations of 1 to 2 g/L are carbohydrates in a chemical sense, but not in a wine-technological sense.

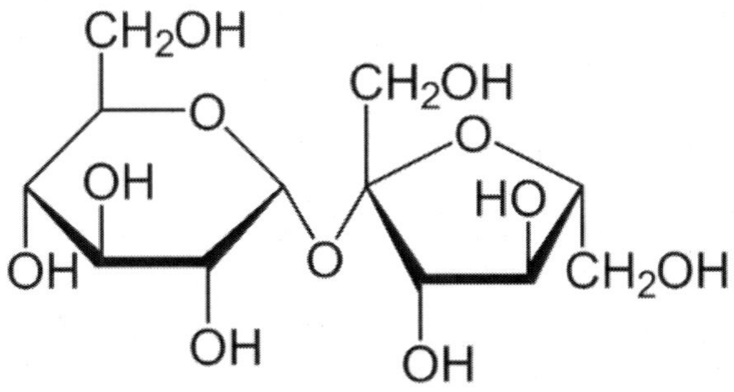

Figure 46. Sucrose structure.

The pentoses that can be found in grapes are usually found in the same quantities in wine. Among pentoses in grapes, arabinose stands out according to its quantity, followed by xylose and rhamnose. The total amount of sugars in grapes, that is, more widely, varies in a fairly wide range and depends on the variety, the conditions of its ripening and the applied measures in cultivation. In most cases, the total sugar content of grapes from cultivated wine varieties ranges between 160 and 250 g/L or between 16 and 25%. Grapes ripened under favourable conditions or left to overripe can contain >30% sugars. The high amount of sugar in the grapes is a prerequisite for obtaining a wine with a high concentration of alcohol or possibly a naturally sweet wine (a wine in which part of the sugar has turned into alcohol, and part gives it sweetness). However, quality wine cannot be obtained from grapes whose only characteristic is high sweetness. The good content of sugars in quality grapes should be accompanied by favourable concentrations of acids and phenolic compounds, but also a developed aroma and a series of other properties that should be translated into an appropriate high-quality wine.

Proper nutrition of the grapevines with potassium has a very favourable effect on the content of sugars in the grapes (Table 24).

It has long been established that potassium increases the rate of transport of sucrose in the phloem and that a significant part of the sugars synthesized in the leaves, during the ripening of grapes, is transferred to the fruit in the form of sucrose (a transport form of sugar) and there it accumulates in form of glucose and fructose (Milosavljević, 1984b).

2.2. Acids in Grapes

The fruits and leaves of the grapevine are rich in organic aliphatic acids. A large number of organic acids are present in grapes, and the most represented are tartaric, malic and citric acids, which represent 96–97% of the total acids (Hidalgo-Togores, 2002). Organic acids in grapes, despite their lower content in relation to sugar, have a significant impact on the organoleptic properties, i.e., the quality of grapes.

It is known that during the ripening of the grapes, the acidity decreases, which happens during oxidation in the respiration process.

Tartaric acid is distinguished according to its concentration, so the content of total (titrating) acids in grapes and wine in all wine-growing and wine-growing countries is expressed in this acid. Tartaric acid follows malic acid in its concentration.

Table 24. Content of total carbohydrates and total acids (g/L) in the must of some grapevine varieties (Stojanova, 2007)

Variety	Total carbohydrates	Total acids
Chardonnay	225	6.40
Italian Riesling	187	6.14
Cardinal	180	4.50
Afus ali	195	4.00
Merlot	250	4.20
Cabernet Sauvignon	105	6.60

In addition to these two acids, grapes always contain up to 1 g/L of citric acid. A series of other organic acids also participate in the formation of the acidity of grapes, the content of which, compared to the three mentioned, is usually significantly lower. Increased content of citric, acetic and gluconic acids can be found in grapes that are attacked by *Botrytis cinerea*. In the formation of the acidity of the wine, in addition to the acids and their salts that are represented in the grapes, the acids that are formed more or less regularly during the alcoholic fermentation, among which acetic and succinic acids should be mentioned in particular, also participate. Additionally, wines that have undergone malolactic fermentation (microbiological transformation of malic to lactic acid) contain significant amounts of lactic acid (Table 24).

The acidity of must and wine is characterized by two indicators: the number of total acids (so-called total or titratable acidity) and real acidity, that is, pH value. The amount of acids in grapes from wine varieties usually ranges from 4 to 14 g/L, and in wines, the amount of acids is from 4 to 8 g/L (expressed as tartaric acid). Wines typically contain less acid than must, because some tartaric and other acids precipitate as salts during alcoholic fermentation and wine aging, forming deposits on the bottom and walls of wine vessels (lids or winestone).

Knowing the must's pH (actual acidity) can also be of great help to winemakers. The sulfurization dose depends on the pH value of the wort. Although a higher total acid content is usually accompanied by a lower pH value, there is not a strong relationship here, since the concentration of hydrogen ions in must or wine is not only influenced by acid ions. The acidity of the grapes, and especially the acidity of the wine, is often considered through the so-called volatile and non-volatile acids (Figure 47). Volatile and non-volatile acids make up the total acids in grapes, must or wine. Most of the total amount of volatile acids in wine is acetic acid.

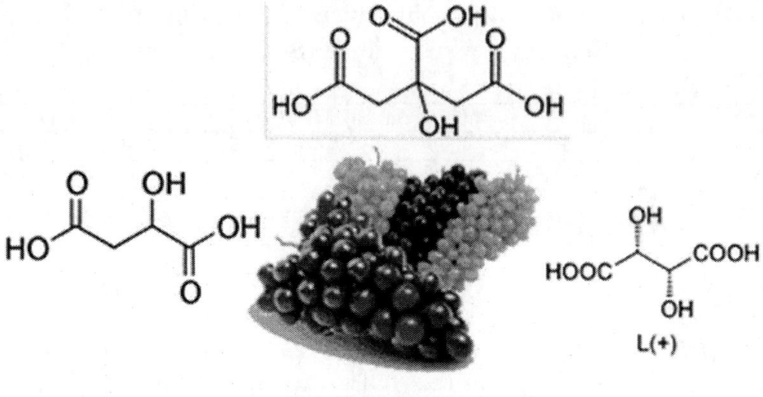

Figure 47. Organic acids in grapevine (Lena do Nascimento Silva et al., 2015).

The appearance of higher concentrations of acetic acid in the wine indicates the beginning or development of some of the spoilage processes of the wine, and the acetic acid gives the wine the typical smell of vinegar. It should be noted that during alcoholic fermentation, yeasts always form smaller amounts of acetic acid, but these amounts of acetic acid do not significantly affect the wine's aroma. The acids present in the grapes have a great influence on the quality of the wine, as well as on its technology. Grapes with a sufficient amount of acids can always be vinified more easily, and the wines obtained from such grapes are easier to stabilize and store. Many varieties of grapevine can bring out a high percentage of sugars with the grapes, which on the other hand is accompanied by a low acid content. In must with a low acid content, the acidity should be increased before the start of alcoholic fermentation. Grape acids, especially tartaric and citric, give the wine a fresh taste. The famous, so-called typical wine taste is related to the taste of the tartaric acid solution. Unlike tartaric acid, malic acid gives the wine a harsh green fruit flavour. In warmer wine-growing regions, malic acid is usually spontaneously converted, and in colder wine-growing regions, under the action of malolactic fermentation, it is converted into lactic acid with a gentler taste for the wine. Citric acid has a great refreshing taste, but its natural content in wines is low, it is microbiologically unstable, so regulations limit the amounts that can be added to the must for increased acidity (Stojanova, 2023).

Nutrition with potassium has a positive effect on increasing the content of total acids in the must and reduces the negative effects of the unfavourable ratio of tartaric and malic acid on the taste of the wine.

2.3. Enzymes

In unripe grape berries, the most important and active enzymes are various invertases, which play a key role in sugar accumulation during ripening. In addition to invertases, dehydrogenases, catalases, oxidases, peroxidases and esterases play a significant role in biochemical activities inside the fruit (Cheynier et al., 2000).

Among the many enzymes found in grapes, due to their influence on wine production, enzymes of redox reactions and hydrolyzing enzymes deserve special attention. Oxido-reduction enzymes can be a problem at the very beginning of grape processing, especially in the production of white wines, because they lead to rapid oxidation characterized by a change in the colour of the must, but also by a change in smell and taste. The action of these enzymes can be counteracted by adding sulfur dioxide. Among the grape hydrolyzing enzymes important for wine production are sucrase (invertase) and pectolytic enzymes. Sucrase participates in the breakdown of sucrose to the fermentable sugars glucose and fructose.

It is found in small concentrations in grapes, and in must significantly larger amounts introduced by dead yeasts. Sucrase is of great practical importance in improving sweetness by adding sucrose, because it breaks it down into glucose and fructose, which yeasts can use. Pectolytic enzymes break down pectin, which is said to make pressing the crushed grapes and later clarifying the wine more difficult. Commercial preparations of pectolytic enzymes have long been used in winemaking. Although the manufacturers of such preparations state that their use has multiple beneficial effects (improving the taste of wine, improving the colour of red wines, etc.), it is certain that through the breakdown of pectin, they contribute to better purification of must, clarification and filtration.

2.3.1. Enzyme Activity during Ripening

Ripening of the fruit is accompanied by an increase in the rate of protein synthesis and the activity of numerous enzymes. Differences in phenolic composition are the result of the activity of enzymes responsible for their biosynthesis and degradation. PAL activity is maximal in very young fruit and then declines during fruit growth. However, in fruits that accumulate anthocyanins, such as grapes, plums, peaches, cherries and strawberries, PAL increases again. The activity of this enzyme can best be illustrated by the example of grapes, where the activity of phenylalanine ammonium lyase is maximal in unripe berries, then decreases during growth, and increases again at the beginning of ripening in red varieties, while it does not change in white varieties (Figure 48). Although PAL activity is necessary for the synthesis of phenolic compounds, it is thought not to be limiting for given physiological conditions. Conversely, its substrate phenylalanine can be a limiting factor in the synthesis of phenolic compounds.

Nevertheless, the comparison of different physiological conditions on numerous examples showed a good correlation between changes in PAL activity and the accumulation of phenolic compounds (in grapes in the case of hydroxycinnamic acid derivatives). In grapes, all studied enzyme activities increase and reach their maximum during the period of greatest anthocyanin accumulation. This includes enzymes specific for flavonoid biosynthesis, chalcone synthase, chalcone isomerase and glucosyl transferase, as well as all enzymes involved in general phenylpropanoid metabolism (PAL, CoAL and C4H). PAL activity is ten times higher in the seeds compared to the skin, while it is negligible in the pulp. Changes in PAL activity are in agreement with changes in the content of anthocyanin in the skin and flavan-3-ol in the seeds (Macheix et al., 1990).

2.4. Mineral Matters

The grapevine absorbs mineral substances from the soil and transports them to all organs, including the grains. In addition to increasing the value of grapes as fruit for consumption, mineral substances have a special role in creating conditions for the alcoholic fermentation of must. The availability of mineral matters greatly affects the activity of wine yeasts and the course of alcoholic fermentation. Mineral matters also reach the wine, where they participate in very complex processes that contribute to the improvement or deterioration of its quality. The amounts of mineral matter in grapes increase during the development and ripening of the berries, although during the ripening period, their accumulation in the berries stagnates. The amount of mineral matter in grapes has a wide amplitude from about 1.5 to >10 g/L, with an average content between 3 and 5 g/L.

The mineral matters in grapes consist of many cations and anions (Figure 49). Among the cations, according to their quantities and significance in wine production, potassium stands out (makes up over 30% of the total minerals in the skin), calcium (in significant amounts in seeds), magnesium, iron, copper and zinc (represented in smaller amounts). Potassium and calcium are the leading minerals in grapes, and potassium alone generally accounts for

>50% of all mineral matter in grapes. Potassium and calcium form salts with the organic acids in the wine (primarily with the tartaric acid), so that significant amounts of them remain in the funnels after pressing or settling in the wine vessels.

In addition to the role of iron, copper and zinc in ensuring the vital functions of the wine yeast, these elements are important for winemaking because, in larger quantities in the wine, they can lead to its different turbidity. The higher amounts of iron in the must, and subsequently in the wine, may be the result of the contact of the grape material or the wine with unprotected iron surfaces. The excessive use of copper-based fungicides, and especially the late treatments of the grapes, can result in the appearance of larger amounts of this element in must and wine.

From the aspect of anions, phosphates, sulfates and chlorides are distinguished according to the amounts in grapes and must. Phosphates are the most abundant anionic mineral substances in the must and due to their role in metabolic processes, but also in wine stabilization processes, they are an important factor for the quality of grapes, must and wine. Grapes usually contain a sufficient amount of microelements important for the development of yeasts and for the synthesis of enzymes (Stojanova, 2023).

During the stages of green berry development and berry ripening, the amount of mineral matter continuously increases. The flow is much faster during the first half of the grape ripening phase than at the end. There is a particularly large increase in potassium.

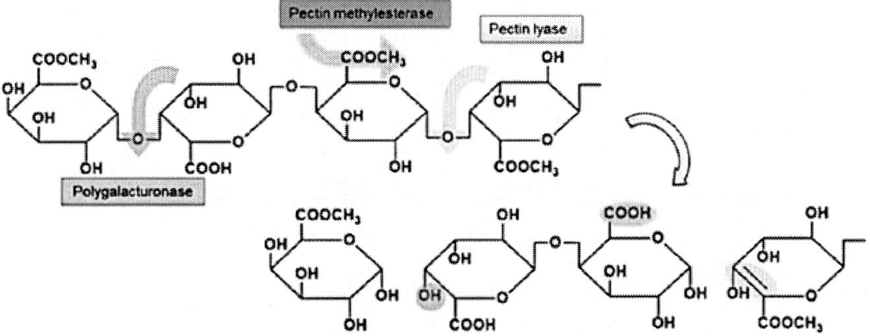

Figure 48. The action of enzymes on grape pectin chains (Scutarașu et al., 2023).

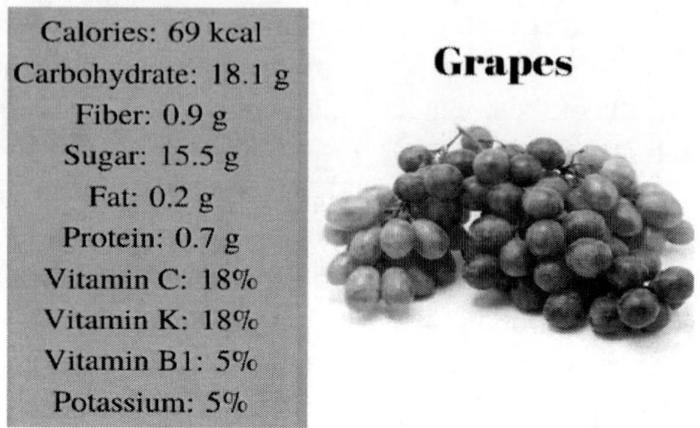

Figure 49. Vitamins and minerals matter in grapes.

2.5. Nitrogen Compounds

Although the amount lags significantly behind sugars, acids and even phenolic compounds, nitrogen compounds in grapes are extremely important for alcoholic fermentation, but also for the quality and stability of wine. Grapes contain almost all groups of organic nitrogen compounds, from proteins, through polypeptides, amino acids, amides, and hexozoamines, to nucleic nitrogen and biological amines.

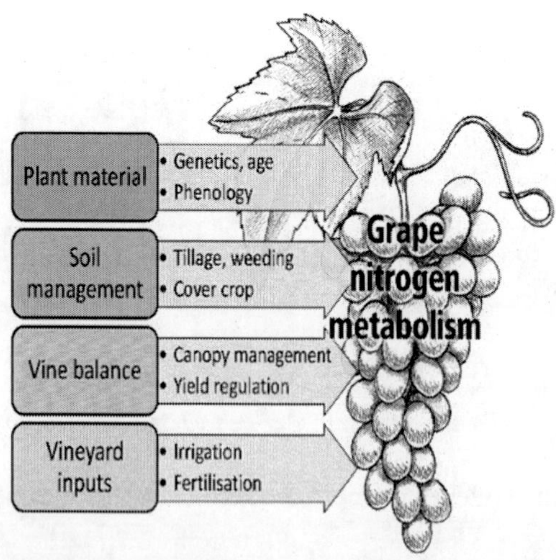

Figure 50. Agronomic practices influencing grape N metabolism (Verdenal et al., 2021).

Free amino acids and soluble proteins belong primarily to this class of grapevine constituents. Among the free amino acids in grapevine fruits, the following were identified: alanine, γ-aminobutyric acid, arginine, glutamic acid, proline, serine and threonine (Stines et al., 2000). Soluble grapevine proteins are broken down by hydrolysis into glycine, alanine, leucine, isoleucine, phenylalanine, tyrosine, threonine and other amino acids.

Due to the specificity of production and the predominant location of nitrogen compounds in the solid parts of the grain, red wines are usually characterized by higher amounts. A significant part of nitrogen compounds in grapes, especially amino acids and inorganic nitrogen, are consumed by yeasts during alcoholic fermentation, so their amounts are regularly lower in wines than in must. Taking into account the importance of nitrogenous compounds for the nutrition of yeasts, and therefore for alcoholic fermentation, by regulating their quantity it is possible to influence this process as well. It is practised to use the removal of nitrogen compounds from sweet wines to biologically stabilize them.

Grapes often do not contain sufficient amounts of nitrogen compounds, especially those used by yeasts as food (ammonia nitrogen, amino acids), so adding nitrogen nutrients to yeasts becomes a common measure (Figure 50).

There are more nitrogen compounds in ripe than in green grapes. There is a particularly large increase from the verasion to the state of full maturity when the content of organic forms of nitrogen (amino acids, polypeptides, peptones and proteins) increases. During the stages of green berry development and berry ripening, the amount of mineral matter continuously increases. The flow is much faster during the first half of the grape ripening phase than at the end. There is a particularly large increase in potassium.

2.6. Lipids

The presence of lipids in grapes is not high, but it is very significant due to the formation of the primary aroma of grapes and wine. They are present in all parts of the berry, and the main lipids are oils, waxes and phospholipids (Figure 51). The seeds are rich in oil, and the waxes are located in the skin and form a wax coating that has a protective role during dry and wet days (Malićanin, 2014). The seeds from grapes (fruit) of the grapevine contain oil, the amount of which varies between 6 and 20% depending on the quality of the raw material and the technological process of obtaining the oil (Baydar et al., 2007; Martinello et al., 2007). The main components of these lipids are free fatty acids: palmitic, stearic, oleic, linoleic and linoleic. The main characteristic of grape seed oil is the high content of unsaturated fatty acids, such as linoleic, 72–76% (Martinello et al., 2007). Besides them, grape seed oil contains sterols, tocopherols (Baydar et al., 2007), as well as phospholipids. This oil is edible and is also used in the cosmetic industry.

Phospholipids are an integral part of cell membranes and play an important structural role, forming micelles and a lipid protective layer. Grape seed oil consists of unsaturated fatty acids (about 90%) and saturated fatty acids (about 10%). Fatty acids are the most important components of fats and oils (acylglycerols) and lipids (phospholipids, waxes, glycolipids, etc.).

2.7. Carotenoids

Carotenoids are present in the pulp and skin of grapes, and although they are not aromatic components in themselves, they are precursors of extremely fragrant compounds that mainly develop during ageing and often result in very high-quality wines. β-Carotene and lutein (85% of the total carotenoids) are usually present, which are liposoluble, and therefore are present only in the juice that has been previously macerated with the skin. Other carotenoids are neochrome, neoxanthin, violaxanthin, luteoxanthin, flavoxanthin, and lutein-5,6-epoxide zeaxanthin (Mendes-Pinto, 2009). Varieties such as Chardonnay and Riesling mostly develop their aroma thanks to carotenoids.

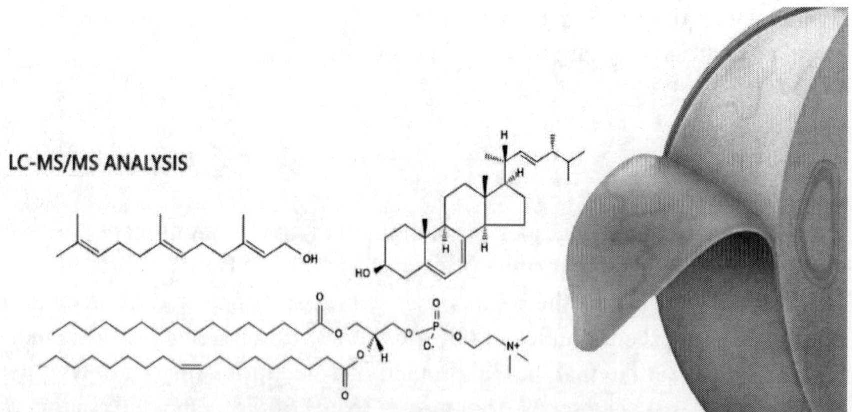

Figure 51. Free fatty acids and other lipids in grapes (Navarro et al., 2019).

2.8. Pectin Substances

Pectin substances usually occur together with cellulose in the intercellular space of plant tissues. Compared to many other fruits, grapes have a relatively low pectin content. Must almost always contains a higher amount of pectin compared to wine. During alcoholic fermentation, pectin breaks down, releasing methanol. The pectin in the solutions acts as a protective colloid, which makes it difficult to carry out technological operations such as must purification, pressing, clarifying and filtering the wine.

2.9. Volatile Components of Grapes That Affect the Aroma

Volatile components of grapes are, in addition to polyphenolic compounds, the most frequently researched class of natural products that can be identified in them. They are mainly responsible for the smell of the fruit (on which the aroma of the wine depends), which is one of the main reasons why so much attention is paid to this class. The components responsible for the aroma of grapes are mainly terpenoids (Koslitz et al., 2008). The most important are o-cymene, limonene, (Z)-β-ocimene, (Z)-linalool oxide, linalool, isoborneol, terpinen-4-ol, α-terpineol, (E)-sabinyl acetate, and (E)-calamenene. In addition to terpenoids, the following grape compounds, norisoprenoids, phenylpropanoids, methoxypyrazines, and volatile sulfur compounds, also influence the aroma of wine (Ebeler & Thorngate, 2009; Gonzalez-Barreiro et al., 2013).

2.10. Vitamins

Grapes are not among the most vitamin-rich fruits, although they regularly contain almost all known vitamins (vitamin C, vitamin B1, vitamin B6). According to the amounts in the grapes, meso-inositol, choline and p-aminobenzoic acid are distinguished. In wine, in addition to vitamins originating from grapes, there are also vitamins originating from yeasts, after their self-degradation (Table 25).

Vitamins are extremely important for the growth and activity of yeasts. Considering that grapes often do not contain sufficient amounts of some vitamins, especially those that are important for yeasts, these vitamins can be found as supplements of nitrogen nutrients for yeasts (Stojanova, 2023).

Table 25. Nutritional value of grapes, juice and wine in 100 g

Component	Grape	Juice	White wine	Rose wine	Red wine
Water (g)	80.54	84.54	86.86	86.40	86.49
Energy (kcal)	69.00	60.00	82.00	83.00	85.00
Proteins (g)	0.72	0.37	0.07	0.36	0.07
Total lipids (g)	0.16	0.13	0.00	0.00	0.00
Carbohydrates (g)	18.10	14.77	2.60	3.80	2.61
Phosphorus (mg)	20.00	14.00	18.00	18.00	23.00
Potassium (mg)	191.00	104.00	71.00	59.00	127.00
Calcium (mg)	10.00	11.00	9.00	10.00	8.00
Magnesium (mg)	7.00	10.00	10.00	10.00	12.00
Iron (mg)	0.36	0.25	0.27	0.20	0.46
Zinc (mg)	0.07	0.07	0.12	0.11	0.14
Copper (mg)	0.13	0.02	0.004	0.005	0.011
Vitamin C (mg)	32.00	25.00	0.00	0.00	0.00
Vitamin B_1 (mg)	0.07	0.02	0,005	0.02	0.005
Vitamin B_2 (mg)	0.07	0.015	0,015	0.015	0.031
Vitamin B_6 (mg)	0.09	0.032	0,05	0.04	0.065
Vitamin E (mg)	0.19	0.00	0,00	0.00	0.00

* Source: USDA FoodData Central, 2021.

2.11. Waxes

Waxes are usually found on the surface of the fruit, in the form of an ashy coating, the thickness of which is variable and depends on the variety. The basic ingredient of waxes is oleanolic acid, which accounts for 50–70%; the rest consists of various higher alcohols, free fatty acids, aldehydes or hydrocarbons.

3. Polyphenolic Compounds in Grapes

Polyphenols are the most studied group of compounds in grapes. They are secondary biomolecules of plants, which include more than 8,000 compounds and are characterized by the presence of at least one aromatic ring with at least one hydroxyl group, free or modified (ether, ester, glycosidic, etc.). The positive effects of grapes, juice and wine on human health are linked to these natural products with proven antioxidant, anti-inflammatory, anti-cancer, neuro- and cardio-protective effects. Also, these compounds are chemotaxonomic markers and have a great influence on the sensory characteristics of juice and wine. They contribute to the final complexity of the wine because their structure and concentration vary depending on the variety, agroclimatic factors, environmental influences and pathogens, as well as the technological production process during which complex reactions and transformations of polyphenols occur (Mitic et al., 2021; Stojanova et al., 2023a; Stojanova et al., 2023b; Stojanova et al., 2024).

Grapes belong to fruits that are rich in polyphenolic compounds and are considered one of the largest sources of these compounds in the human diet. The most abundant classes are anthocyanins, flavan-3-ols, flavonols, phenolic acids and stilbenes and are distributed differently in the berry. The highest content of polyphenols is in the seed, which is characterized by flavan-3-ols of low molecular weight, and procyanidin oligomers that contribute to the bitter taste. The skin is rich in condensed tannins, monomeric flavan-3-ols and flavonols, phenolic acids, resveratrol and anthocyanins. The concentration of polyphenols is the lowest in the fleshy part of the berry. Anthocyanins are the main group of phenolics in red grapes, while flavan-3-ols are characteristic of white varieties. Since the degradation of cell wall polysaccharides is the most important step in the release of phenols from the skin of grapes and their transition into juice and wine, red wines have a higher phenolic content than white wines due to prolonged contact between juice and skin during maceration. Also, certain phenols in wine originate from wooden barrels (Fernández et al., 2015; Hornedo-Ortega et al., 2020; Stojanova, 2023c; Stojanova et al., 2024).

3.1. Phenolic Compounds

Phenols are compounds that have one or more hydroxyl groups directly attached to an aromatic ring. Phenol represents the basic structure of this class of chemical compounds. Phenols are similar in many ways to aliphatic alcohols in which the hydroxyl group is attached to an acyclic hydrocarbon chain. However, the hydroxyl group of phenol is under the influence of the aromatic ring, which causes weaker hydrogen bonding, making phenols mild acids (Vermerris & Nicholson, 2006).

Phenolic compounds are found in large numbers in plants and are products of their secondary metabolism. They are usually found in plant tissues in the form of glycosides or esters, rather than free compounds (Vermerris & Nicholson, 2006). Phenolic compounds comprise a very large and diverse group of molecules, with over 8,000 structures characterized so far, ranging from simple molecules such as phenolic acids to highly polymerized compounds such as tannins. These compounds can be classified in several ways. Harborne & Simmonds (1964) propose a classification based on the number of carbon atoms in the molecule. Alternative classifications were proposed by Swain & Bate-Smith (1962), according to which phenolic compounds were classified into "common" and "less common" according to their occurrence in nature. Ribéreau-Gayon (1972) grouped phenolic compounds into three classes as follows:

1. Widespread phenols, present in all plant organisms or of specific interest for a certain plant species;
2. Phenols that are less abundant in nature, with limited by the number of known structures;
3. Phenolic compounds that occur in the form of polymers.

Natural phenolic compounds have been the centre of interest of many research groups in recent decades since they are found in significant quantities in food products of plant origin, and there are indications supported by clinical studies that their regular consumption reduces the risk of numerous diseases (Ramos, 2008). Phenols are widespread constituents of plant foods (fruits, vegetables, cereals, vegetable oils, chocolate, etc.) as well as beverages (fruit juices, coffee, beer, wine, spirits, etc.) and are partly responsible for the formation of their overall organoleptic properties. For example, phenols contribute to the degree of bitterness and astringency of fruit juices due to the interaction of compounds from the procyanidin class with salivary glycoproteins. Anthocyanins belonging to the flavonoid group are responsible for the orange, red, blue and purple colours of fruits and vegetables. It is also known that phenols such as vanillin, syringin, etc. are responsible for the formation of the aroma, specific bouquet and colour of the wine, as well as many other alcoholic drinks aged in wooden vessels. Many phenolic compounds, especially those from the class of flavonoids, have been shown to possess a wide range of biological activities (Djukic & Jemcev, 2003; Djukic et al., 2015) such as antioxidant, antitumor, antihypertensive and antidiabetic. Recently, more and more attention has been paid to the study of other classes of phenolic compounds, such as phenolic acids and tannins, which were once thought to have no significant biological activities (Robbins, 2003). The selection of extraction solvents is critical, as it will determine the amount and type of phenolic compounds being extracted (Boskovic et al., 2022; Stojanova et al., 2021; Stojanova et al., 2024).

Table 26. Phenolic compounds in (whole seed, seed and leaves) of red grape (sultana cultivar) (mg/100g) (Hussein et al., 2015)

	Whole seed	Seed	Leaves
Vanillic acid	1.45	19.85	0.85
Catechin	779.57	1799.29	44.08
Protocachoic	8729.55	210.00	811.23
Coumarin	11.89	216.27	2.22
Gallic	889.20	2823.03	44.93
Ferulic	13.00	33.42	59.90
Catechol	5533.14	33.90	32.90
Chlorogenic	4039.26	3903.40	199.43
Synergic	440.30	19.50	17.80
Pyrogallol	58.68	45.23	139.75
Caffiec	7.25	5.33	6.69

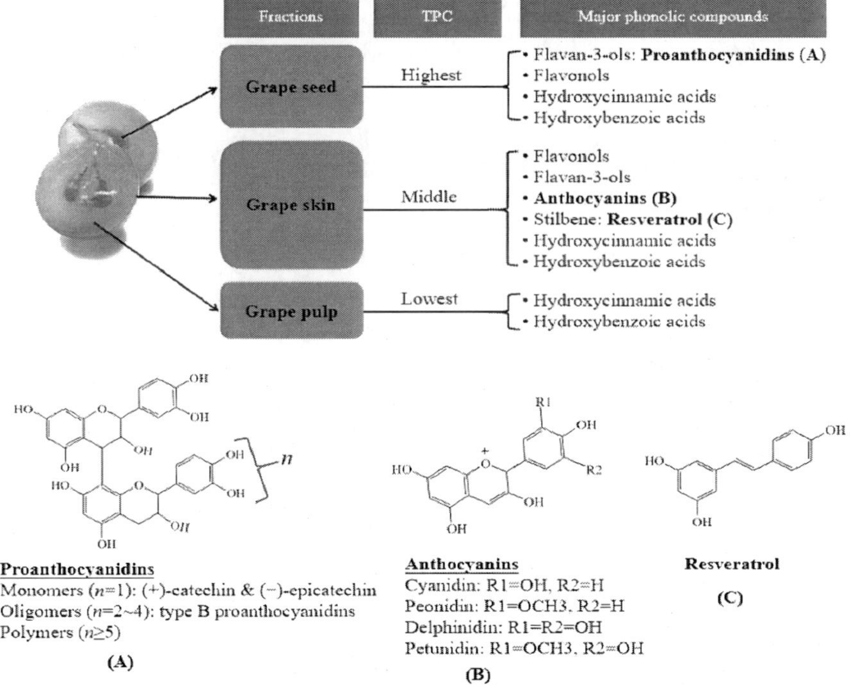

Figure 52. Phenolic compounds in different fractions of grape and the chemical structures of several representative bioactive compounds. (A) Proanthocyanidins; (B) anthocyanins; (C) resveratrol. TPC, total phenolic content (Zhou et al., 2022).

Class	Basic skeleton	Basic structure
Simple phenols	C_6	
Benzoquinones	C_6	
Phenolic acids	C_6-C_1	
Acetophenones	C_6-C_2	
Phenylacetic acids	C_6-C_2	
Hydroxycinnamic acids	C_6-C_3	
Phenylpropenes	C_6-C_3	
Coumarins, isocoumarins	C_6-C_3	
Chromones	C_6-C_3	
Naphthoquinones	C_6-C_4	
Xanthones	C_6-C_1-C_6	
Stilbenes	C_6-C_2-C_6	
Anthraquinones	C_6-C_2-C_6	
Flavonoids	C_6-C_3-C_6	
Lignans and neolignans	$(C_6$-$C_3)_2$	
Lignins	$(C_6$-$C_3)_n$	Highly crosslinked aromatic polymer

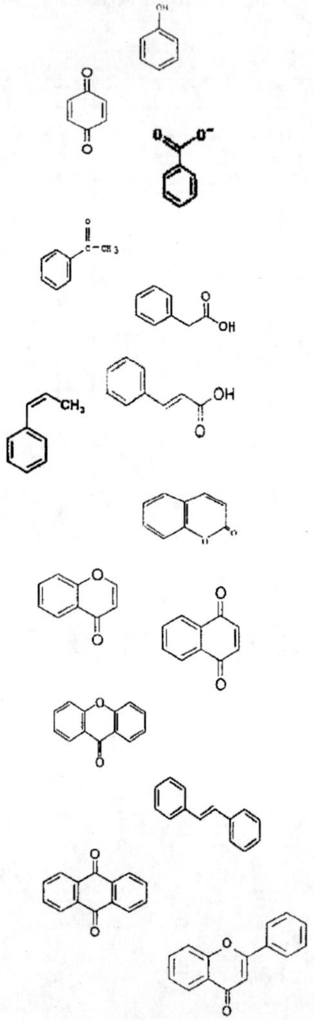

Figure 53. Main classes of phenolic compounds regarding their carbon chain (Reis Giada, 2013).

Phenolic compounds are one of the most important compounds in grapes intended for processing into wine. Phenolic compounds are responsible for the colour of the wine, their astringency, and possibly for the bitterness and play a major role in creating pleasant but also unpleasant aromas of the wine stability and age-ability of wine. The reactions in which phenols participate in the material from grapes and wine are influenced by several factors (conditions of the external environment, applied agricultural techniques and varieties), so they go through a series of changes in their state almost non-stop. Grapes and wine contain a large number of phenolic compounds that undergo significant changes during the processing of grapes, ripening and ageing of the wine, bind to each other, and change colour or precipitate. The skin colour of varieties is different even if they are grown under the same external conditions, so some varieties have pink, others red or bluish skin colour (Pržić, 2015).

Examining the content of phenols in the skin and mesocarp of the berry is of great importance because they influence the plant itself, but also because of their positive effect on human health. The content of phenolic compounds in grapes mostly depends on the variety (Table 26), but climatic conditions, applied agrotechnical measures, soil composition, ripeness of grapes, etc. have a great influence on their composition and quantity (De Pascali et al., 2014; Cheng et al., 2015).

Of the total content of phenolic compounds in grapes, about 60% is found in the seeds, about 20% in the stem, and about 15–20% in the skin. Proanthocyanidins or flavan-3-ols make up 50–70% of the phenolic compounds of the seeds contain 28 flavanol units and have pronounced tannic properties (Hayasaka et al., 2003). In the seeds, catechin, epicatechin and gallocatechin are most present in the esterified form with gallic acid. A high content of these compounds occurs after fertilization and implantation of the berry. As a result of the effect of light, the genes for tannin synthesis are inactivated in the skin, and the genes for tannin synthesis are activated in the seeds (Mattivi et al., 2009). As the seeds mature, from the hinge towards full maturity, the dark colour of the seed coat appears due to the polymerization and oxidation of tannins. The tannins of the seed appear as a barrier between the tannins concentrated in the seed and prevent the transfer of the entire amount of tannins into the wine.

Bindon et al., (2011) found that phenolics in different parts of grapes respond differently to ripening. They recorded a decrease in tannin in the seed with ripening, and an increase in phenol in the skin was also observed. Many interrelated factors play a role in determining the composition and concentration of phenols in red grapes and wine. The response of different varieties to these factors is very specific. Although some factors can be manipulated during grape and wine production, others, such as the climate, are difficult to control. The content of anthocyanins and proanthocyanidol extracted into wine depends on their quantity in the grapes. At the beginning of the maceration process, the phenolic components are most easily extracted from the skin (Puškaš, 2010). However, their content in the skin is significantly lower compared to seeds and spores, so seeds and spores are the main sources of these compounds in wine.

Differences in the metabolism of phenolic compounds, from the aspect of physiological conditions, depend directly on protein synthesis. However, secondary metabolism in general, and especially the metabolism of phenolic compounds, also depends a lot on numerous factors, either external (light, temperature, water supply, fertilization, etc.) or internal (hormones, etc.). Knowing the role of these factors in regulating the metabolism of phenolic compounds is particularly interesting due to the possibility of practical application in terms of intensification, i.e., increasing or decreasing the synthesis of certain groups of phenolic compounds in the fruit. A decrease in flavonoid biosynthesis was observed when either endogenous or exogenous factors were of low or high intensity of action (Braidot et al., 2008).

Phenolic compounds are distributed in all organs of the grapevine (Figure 52). Grapevine leaves are rich in flavonols, flavan-3-ols, phenolic acids and stilbenes (Pezet et al., 2003; Dani et al., 2010). The petiole is rich in flavonols, flavan-3-ols and phenolic acids. The fruit is rich in different classes of phenolic compounds that are differently distributed in the skin, seeds and pulp of the berry. The seeds are rich in flavan-3-ol monomers, oligomers and gallic acid derivatives (Kennedy, 2008). The composition of the skin differs in white and red grapevine varieties. Red varieties contain anthocyanins, flavonols, flavan-3-ols, phenolic acids and stilbenes in their skin, while white varieties have no anthocyanins and other classes of phenolic compounds are present in larger quantities. The pulp contains a small amount of these compounds, mainly phenolic acids (Pastrana-Bonilla et al., 2003). There are many classifications of phenolic compounds (Pereira et al., 2009). Based on the chemical structure, the following have been proposed:

- phenolic compounds of the non-flavonoid structure;
- flavonoids (flavonols, flavan-3-ols, procyanidins, anthocyanins, tannins, etc.).

Phenolic non-flavonoid compounds are mainly accumulated in the mesocarp of the berry, while flavonoids are concentrated in the skin, seeds and pulp. Phenolic compounds of the non-flavonoid structure are phenolic acids and stilbenes and contain C6-C1, C6-C3 or C6-C2-C6 in the basic skeleton.

3.1.1. Phenolic Acids

Phenolic acids are present in all organs of the grapevine in different amounts. Phenolic acids are hydroxy and methoxy derivatives of benzoic and cinnamic acids. Grapes and wine contain benzoic acid derivatives: gallic, p-hydroxybenzoic, protocatechinic, vanillic, syringitic and ellagic acids. Among the cinnamic acid derivatives are present: trans-p-coumaric, trans-caffeic, trans-ferulic and chlorogenic acid, and in the form of cinnamic and tartaric acid esters are present: trans-caftaric, cis- and trans-cutaric, trans-fertaric and sinapic acid (Figure 53).

Phenolic acids in grapes are present in the glucosidic form, from which they can be released by acid hydrolysis, and in the form of esters, from which they are released by alkaline hydrolysis. Hydroxycinnamic acids are partly responsible for the colour of red wines because their oxidation produces yellow-coloured compounds. The antioxidant activity of phenolic acids depends on the number and position of hydroxyl groups in the benzene nucleus. Therefore, gallic acid, which has three hydroxyl groups, is considered the best phenolic antioxidant. The most significant biological action of phenolic acids is their influence on lipid peroxidation, which is why they can be used as preservatives and anti-inflammatory agents (Cushnie & Lamb, 2011; Tian et al., 2009). A smaller part of phenolic acids is found in free form, while the majority is bound by ester, ether or acetal bonds to the structural components of the plant (cellulose, proteins and lignin) or smaller organic molecules such as glucose, maleic and tartaric acid (Zhu et al., 2012).

3.1.2. Stilbenes

Stilbenes are compounds that under normal circumstances cannot be found in the organs of the grapevine, or are present in a very small amount (Pezet et al., 2003; Anastasiadi et al., 2010). These ingredients, under the influence of various environmental factors, are rapidly synthesized and accumulated in its tissues, and are collectively known as “phytoalexins.” These are compounds with strong antimicrobial activity, which the plant synthesizes in response to bacterial and fungal infections and are considered the main bearers of the resistance mechanism against various plant diseases. Among these compounds, stilbene derivatives take an important place: resveratrol, piceid and viniferins (oligomers of resveratrol), cis and trans isomers of these compounds.

Stilbenes are phenolic compounds in which two aromatic nuclei are bridged by an ethane or ethene bridge. Stilbenes (1,2-diarylethenes) are non-flavonoids located in the skin, but also in significantly smaller amounts in the seeds. Significant amounts of these compounds are present in the bunch of grapes. Resveratrol (monomeric stilbene) is the main stilbene in grapes. Stilbenes present in oligomeric and polymeric forms are known as viniferins. Viniferins identified in grapes are: α-, ε-, δ-viniferin, dimeric and trimeric resveratrol (Ali et al., 2010).

3.2. Flavonoids

Flavonoids represent the largest and most important group of phenolic compounds. The most important representatives of this group of compounds are tannic substances and coloured substances of grapes and wine. Flavonoids include all phenolic compounds that have the common structural formula C6-C3-C6, such as catechins, dihydrochalcones, chalcones, flavanones, flavones, flavanonols, flavonols, anthocyanidins, leucoanthocyanidins and aurones (Keller, 2010).

Flavonoids are present in the skin and seeds. According to their structure, they are compounds with 15 atoms arranged in the basic structural formula according to the C6-C3-C6 system, where nine atoms belong to the benzopyran ring (benzene ring A which is condensed with pyran ring C). The remaining six atoms form the benzene ring B which can be connected to the benzopyran ring at positions two, three and four. Linking at position two results in flavones, flavonols, flavonones, dihydroflavones, flavan-3-ols, flavan-3,4-diols and anthocyanidins, position three: isoflavones and position four: neoflavones. By changing the groups in positions two, three and four, as well as introducing other substituents (sugars), condensed forms of tannins, glycosides, etc. are built (Keller, 2010).

The flavonoid composition of the skin is similar in both red and white grapes (Pinelo et al., 2006; Braidot et al., 2008), and as is known, it varies during grape ripening. The accumulation of proanthocyanidins occurs from the stage of fruit formation until 1–2 weeks before the hinge, and then their concentration decreases between the hinge and harvest. Unlike tannins, the accumulation of anthocyanins begins at the hinge and reaches a maximum in the last stages of fruit ripening, although there may be a drop in concentration, especially in warm climates. Flavonol synthesis occurs in two distinct periods, the first near flowering and the second starting after the hinge (Downey et al., 2006). However, their total concentration is higher at the beginning of fruit development.

Flavonoids form a large subgroup of phenolic compounds that represent the most important group of compounds that enter the chemical composition of the berry. Most phenolic compounds have important physiological functions, while some of the phenols can be recognized as compounds with vitamin properties (vitamin P). Phenolic compounds affect the organoleptic properties of grapes, must, and grape juice, but they also contribute to better and longer wine storage and have a positive effect on the wine ageing process.

Numerous studies have shown that flavonoids are antioxidants, good scavengers of free radicals, chelators of metal ions and have significant biological properties (anti-inflammatory, anti-allergic, anti-cancer, anti-hypertensive and antimicrobial) (Youdim et al., 2002).

The composition of flavonoids changes during seed maturation, along with macroscopic changes in tissues, such as colour and hardness (Figure 54). Before the verasion, the inner layers of the middle sheath undergo a process of thickening and strong lignification of the cell walls, while the cells of the soft outer sheath begin to intensively acquire colour after the hinge. It is believed that the darkening of the seed, during ripening, is mainly the result of the oxidation of flavan-3-ol and tannins accumulated in the thin-walled cells of the outer coat, which provides a physical and chemical barrier to oxygen and infection by pathogens.

The highest concentration of flavan-3-ols (monomers and condensed tannins) was recorded in the verasion when the accumulation reached a maximum, and then during ripening, it decreased by 90% in the case of monomers and 60% for proanthocyanidins (Braidot et al., 2008).

3.3. Flavonols

Flavonols are a very important class of phenolic compounds that are found in grapevines in leaves, stems and skin, but are not present in seeds and pulp. The most common flavonols in grapevines are quercetin, rutin, morin, myricetin and kaempferol, and in addition to the free form, they are most often found in the form of glucosides, galactosides and rutinosides (Pastrana-Bonilla et al., 2003; Monagas et al., 2006; Dani et al., 2010).

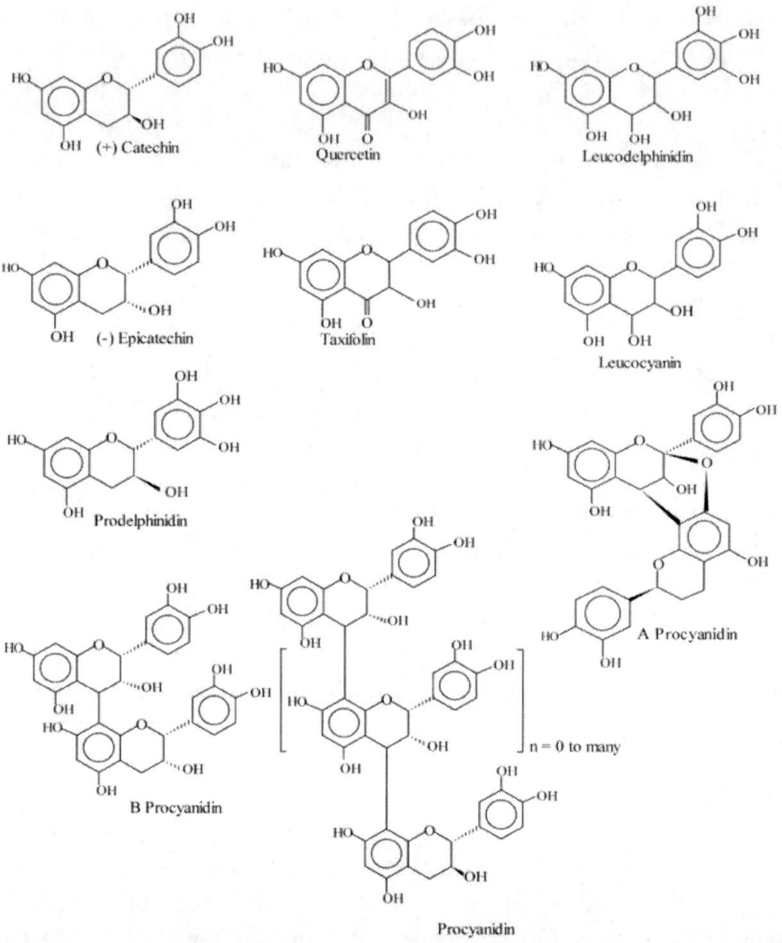

Figure 54. Flavonoids present in grape skin, grape seed and pine bark extracts (Wood et al., 2002).

3.4. Flavan-3-ols

Flavan-3-ols are the most abundant class of phenolic compounds found in grape seeds, but they are also present in skin, petioles and leaves of grapevine (Anastasiadi et al., 2010), leaf petioles, leaves, in the form of monomers, dimers, oligomers and polymers (Zhu et al., 2012). The following were isolated from the extracts of various grapevine organs: (+)-catechin, (-)-epicatechin, (+)-gallocatechin, (-)-epigallocatechin and (-)-epigallocatechin gallate (Anastasiadi et al., 2010; Zhu et al., 2012). Oligomeric and polymeric derivatives of flavan-3-ols known as condensed tannins or proanthocyanidins were also found in significant amounts in grapevines (Anastasiadi et al., 2010; Lachman et al., 2009). They consist of multiple monomeric units interconnected by interflavanic bonds, most often of (+)-catechin and (-)-epicatechin. Proanthocyanidins are composed of terminal and extended flavan-3-ol monomeric units, interconnected by C-4 interflavanic bonds and C-8, as well as the less abundant C-4 and C-6. In grape skin, catechin is the primary terminal unit associated with epicatechin and epicatechin gallate to a much lesser extent. Epicatechin is the most common extension unit, followed by epigallocatechin and epicatechin gallate (Downey et al., 2003; Monagas et al., 2003).

3.5. Proanthocyanidins

Proanthocyanidins are composed of the terminal and extended flavan-3-ol monomer units, interconnected by interflavan bonds C-4 and C-8, as well as less frequently C-4 and C-6. In grape skin, catechin is the primary terminal unit associated with epicatechin and epicatechin gallate to a much lesser extent. Epicatechin is the most common extension unit, followed by epigallocatechin and epicatechin gallate (Downey et al., 2003; Monagas et al., 2003). The most important representatives of proanthocyanidins are proanthocyanidin A1, proanthocyanidin B2 and proanthocyanidin C2. Grapevine variety also affects the degree of polymerization of proanthocyanidins (Mattivi et al., 2009), which affects the extraction from the skin and seeds during the winemaking process (Gambuti et al., 2009). In general, skin proanthocyanidins are extracted earlier during the fermentation process and with increasing maceration time, the extraction of seed proanthocyanidins increases. The evaluation of the polyphenolic potential enables the determination of differences in the composition of polyphenols and provides factors with which to evaluate the oenological potential of grapes.

3.6. Anthocyanins

Anthocyanins (Greek: *anthos*-flower, *kyanos*-blue) are plant pigments of red, purple or blue colour. According to their chemical structure, anthocyanins are anthocyanidin glycosides, so anthocyanin hydrolysis results in anthocyanidins (aglycones) and the remaining mono or disaccharides. Anthocyanidins are structural derivatives of flavylium cations (Harborne & Williams, 2001). All anthocyanidins have a hydroxyl group on the C3 atom, and the other groups are distributed on C5, C7, C3', C4', and C5'. One of the most important characteristics of anthocyanins is that the oxygen in the heterocyclic ring can have a positive charge, and thanks to this, they act as cations in an acidic environment and build salts with acids (they have a red colour), and in an alkaline environment as anions and build salts with bases (have blue colour).

The colour of anthocyanin depends on the number of hydroxyl groups (higher number of OH-groups - blue colour, lower number - red colour). Methylation of the hydroxyl groups on ring B increases the colour intensity, so pelargonidin has an orange-red colour, cyanidin - dark red, and delphinidin - purple. Of the sugars, anthocyanins contain glucose, galactose and rhamnose. If an anthocyanin contains one sugar, it is most often found in position 3 (Figure 55). Anthocyanins are responsible for the colour of red grapes and wine. In grapes, they are found in the skin and pulp of the bunch. They are found in the form of monoglycosides in the grape variety *Vitis vinifera*, while in the form of diglycosides in the American varieties. It is known that the content and type of anthocyanin depend on the grape variety, the conditions of growing the grapevines, the maceration process of the grapes and the way the wine is stored. So the colour of the wine obtained from the same grape variety can be different depending on the geographical area of cultivation, processing technology (temperature and duration of maceration, amount of sulfur (IV) oxide, etc.) and the way of storing the wine.

Light plays a crucial role in the synthesis of anthocyanins, namely wines from southern wine-growing areas are more coloured compared to wines from northern wine-growing areas (Figure 56). The composition of the anthocyanin complex and the concentration of certain groups of anthocyanins are influenced by the ecological conditions of the locality (Marković, 2012). Grapevine variety also affects the degree of polymerization of proanthocyanidins (Chira et al., 2009), which affects the extraction from the skin and seeds during the winemaking process (Gambuti et al., 2009). In general, skin proanthocyanidins are extracted earlier during the fermentation process and with increasing maceration time, the extraction of seed proanthocyanidins increases.

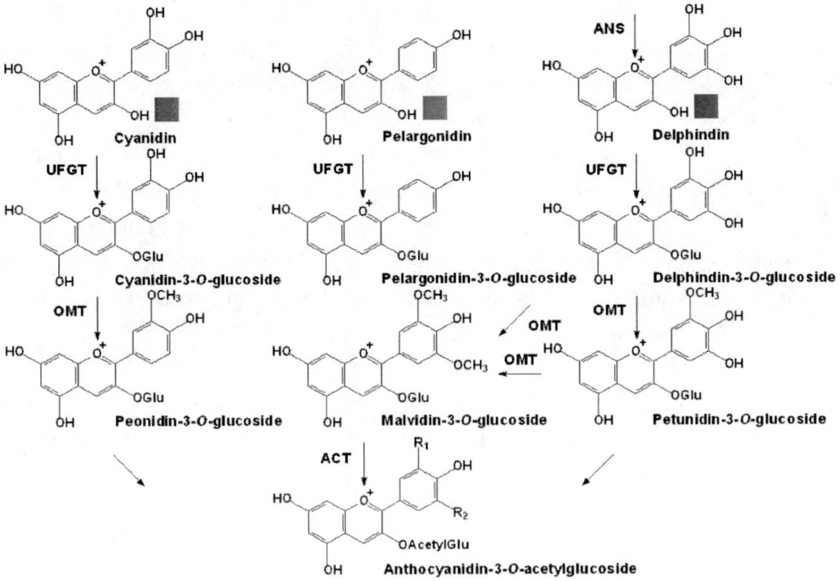

Figure 55. The specific pathway for the anthocyanin modification of free anthocyanidins in grapes. UFGT, flavonoid glucosyltransferase; OMT, O-methyltransferase; ACT, anthocyanin acyltransferase (He et al., 2010).

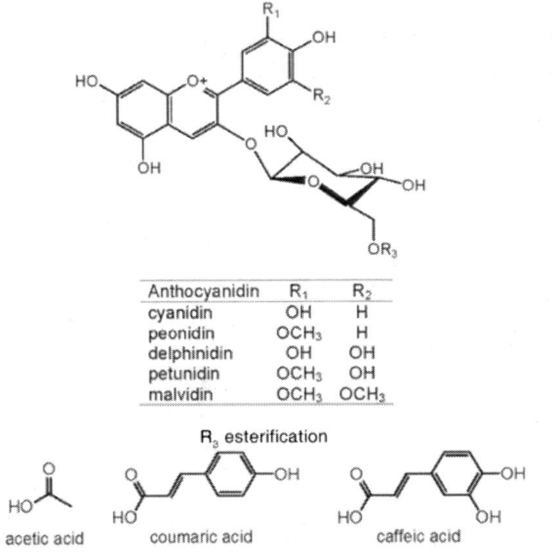

Anthocyanidin	R_1	R_2
cyanidin	OH	H
peonidin	OCH_3	H
delphinidin	OH	OH
petunidin	OCH_3	OH
malvidin	OCH_3	OCH_3

R_3 esterification

Figure 56. Structures of grape-derived anthocyanins (Kennedy et al., 2006).

The use of foliar fertilizers (especially P and K fertilizers) indicates a positive effect on the accumulation of anthocyanins. This is a result of the crucial importance of phosphorus and potassium for the synthesis of carbohydrates and the transport of assimilates to the storage organs, and because the factors that affect the content of sugars also affect the colour of the grapes, which depends on anthocyanins (Champagnol, 1984). He et al., (2010) state two different opinions on the relationship between anthocyanins and sugars. According to the first, sugars in the epidermis have a regulatory role in anthocyanin biosynthesis; while according to another opinion, they are important only as substrates for the formation of anthocyanins.

Anthocyanins which are mainly found in grapes as glucosides, are responsible for the colour of red juice and wine. In that form, they are chemically more stable and soluble in water, and they can also be bound with acetic, coumarin and caffeic acid. The colour largely depends on their structure, which changes with pH, so in a more acidic environment, a red colour stands out, while at a higher pH, shades of blue-violet colour come to the fore. Also, the shade of colour depends on the degree of hydroxylation of the B-ring, as well as the ratio of free and methylated hydroxyl groups, and the presence of SO_2 can lead to bleaching. Malvin is the dominant anthocyanin in the reddest grape varieties, and since it is the reddest of all anthocyanins, it is responsible for the colour of young red wines. Anthocyanins, which in grapes are mostly connected by hydrophobic bonds, or to other phenols (this phenomenon is called co-pigmentation), dissociate from these aggregates during vinification and wine maturation and can lose their sugar and acyl components, which makes them very sensitive to oxidation that leads to the darkening of the colour of the wine. That is why sufficient amounts of catechins, proanthocyanidins and condensed tannins must be present in the wine, which will polymerize with the anthocyanins, thus stabilizing the colour and increasing its intensity. Anthocyanins and flavonols are localized in the skin. At the same time, flavan-3-ols and procyanidins are mostly synthesized in the seeds and stems and to a lesser extent in the skin, and can polymerise already at the beginning of vinification (Stojanova et al., 2024).

Categorization of grapevine varieties can also be done based on the degree of accumulation and concentration of anthocyanins (varieties with a higher concentration, varieties with medium and varieties with a lower concentration of anthocyanins in the skin of the berry).

After the process of fertilization and setting of the berries, that is, during the warmest part of the vegetation, the biosynthesis of tannins begins, and with the entry into the phenophase of the berry, the synthesis of tannins and their polymerization stops. Parallel to the cessation of tannin synthesis, the synthesis and increase of anthocyanin content in the skin of the berry begins (Harbertson et al., 2002; Hanlin & Downey, 2009). Regardless of the different speed and concentration of tannin accumulation in relation to the phenophase in which the berry is, the percentage of tannin extraction into wine is almost unchanged, that is, it does not follow the speed of accumulation in the berry. 25–75% of tannins are extracted from the berry, of which 50–80% originate from the skin, the degree of their extraction depends on the ripeness of the berry, the duration of maceration and the applied technological measures during wine production. Liu et al., (2010), point out that with an increase in the percentage of alcohol, the extractability of grape phenols into wine increases. This may be due to the degradation of the cell wall, which is caused by the increased alcohol content, and therefore the increase in phenol extractability in the skins of ripe grapes.

Bindon & Kennedy (2011) found that polymeric proanthocyanidins in wine are released in higher concentrations from ripe grapes. This is probably because proanthocyanins are less reactive with other phenols, such as anthocyanins in berry skins.

Regarding the influence of nutrients added by fertilizing, many studies have shown an increased content of anthocyanin in the case of the application of moderate doses of nitrogen (50 g/plant) and, as already mentioned, a decrease in the content of solid soluble substances (Delgado et al., 2004). In addition, an increase in anthocyanin content by 23% and 40% was recorded when applying 60 and 120 g of potassium per bunch, respectively (Delgado et al., 2004).

Excessive application of nitrogen and potassium fertilizers leads to more vigorous growth, which delays ripening and reduces the colour of the grapes. Such effects are a consequence of the induction of metabolic imbalance and competition between vegetative parts and berries for sugar translocation. Due to greater lushness, the exposure of grapes to the sun is limited, which together with temperature represents the main environmental factors that can affect the biosynthesis of flavonoids (Delgado et al., 2004). Optimum amounts of nitrogen (especially in the form of urea) have a positive effect, and higher amounts, especially in the second half of the vegetation, have a negative effect.

Potassium nutrition is very important for colour formation in cultivars characterized by reddish bunches, as it affects the synthesis of anthocyanins and polyphenols. Potassium stimulates photosynthetic activity and affects the translocation of sugars in the bunch, thus indirectly affecting the synthesis of phenolic compounds. Optimal magnesium nutrition also has a positive impact on wine quality. It encourages the formation of substances that give the wine a characteristic bouquet. In case of its deficiency, the varietal characteristics of the wine are less pronounced (Stojanova, 2023).

3.7. Tannins

Tannins represent the most diverse group of flavonoid compounds with astringent and pungent tastes. According to the chemical classification, there are two subgroups of tannins:

1. Water-soluble (hydrosoluble);
2. Condensed.

Water-soluble tannins mostly consist of gallic, ellagic acid molecules of glucose. In contrast, condensed tannins are mostly flavan3-ols, such as catechin and epicatechin. Catechin and epicatechin in their monomeric forms impart a bitter taste (Dixon et al., 2005). Only condensed tannins are present in grapes, while water-soluble tannins are found in wines.

Water-soluble tannins come from the wood used in the cellar vessels in which the wine often spends shorter or longer periods during maturation, processing and storage, and the main sources of condensed tannins are grape skins and seeds. Although water-soluble and condensed tannins differ according to their chemical structures, both are classified as tannins, primarily due to their ability to bind and precipitate proteins.

Condensed tannins (Figure 57) are also called protoanthocyanidins, which in their chemical composition are close to anthocyanins, they are colourless or yellow, and in an acidic environment when heated, they turn into red-coloured anthocyanidins, where the enzymes leucocyanidin reductase and anthocyanidin reductase play a key role in this process. In other words, depending on the enzyme and the biochemical cycle, anthocyanidins can be converted into tannins or anthocyanins. Under these conditions, leucoanthocyanins change to cyanidins (Bogs et al., 2005). With the start of fermentation of the beak and extraction, tannins and anthocyanins can polymerize with each other, which results in greater stability of the wine's colour. During the polymerization of tannins and anthocyanins in wine, bitter flavours, a certain amount of acridity, a dry impression, but also the smell of roses are defined (Gawel, 1998).

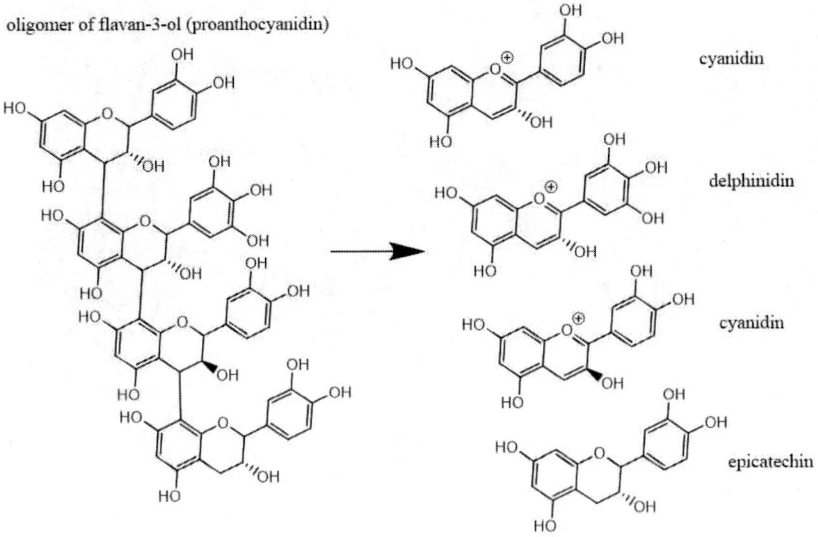

oligomer of flavan-3-ol (proanthocyanidin)

cyanidin

delphinidin

cyanidin

epicatechin

Figure 57. Condensed tannins or proanthocyanidins releasing anthocyanidins after an acid treatment with alcohol and high temperature. On the left, the molecule of tannin is composed of epicatechin, epigallocatechin, catechin and epicatechin (from top to bottom), and it releases different anthocyanidins (on the right) (Iowa State University, 2020).

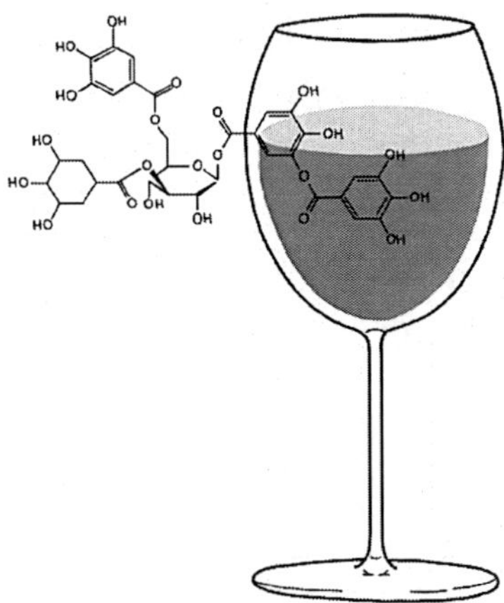

Figure 58. Tannins in wine.

Figure 59. Interactions of wine polysaccharides with aroma compounds, tannins, and proteins (Jones-Moore et al., 2022).

Tannins from grape skins and grape seeds differ from each other and have different effects on wine characteristics. As a rule, tannins from the skin have a higher degree of condensation (polymerization), and are characterized by astringency, but not bitterness, which is why they are called "good tannins." Tannins from the seeds have a lower molecular weight and often contribute to the bitterness of wines, which is why they are called "bad tannins." It should be emphasized that during the production of red wines that are consumed as young wines or within two to three years after harvesting the grapes, maceration is necessary to extract sufficient amounts of anthocyanins that will give these wines the appropriate colour. However, in the production of high-quality red wines that will age in bottles, with longer or more intense maceration, it is first necessary to extract larger amounts of tannins that give old red wines the necessary structure (Stojanova, 2023).

The ability to bind tannins with cell wall pectins and glucans increases significantly immediately after the hinge, but before full maturity with the breakdown of most of the cell wall pectins, the ability to bind tannins significantly decreases (Kraus et al., 2003). According to Fournand et al., (2006) despite the different speeds and concentrations of tannin accumulation depending on the phenophase in which the berry is located, the degree of tannin extraction into wine is almost unchanged, i.e., it does not follow the rate of accumulation in the berry. Depending on the degree of ripeness of the berry, the length of maceration and the wine-making technology, between 25–75% of tannins are extracted from the berry, of which 5–80% comes from the skin. After separation from the wine, the remaining fermented leaves remain enriched for a part of non-extracted substances, which qualifies it as such for use for other purposes. From such residual juice, a high-quality brandy known as "grappa" can be made in the distillation process. The highly valued grape seed oil can be obtained by pressing the remaining seeds. The beak can also be used for pharmaceutical purposes for extracting the basic raw materials of herbal medicines or for the production of dietary preparations. Finally, such a hook can be used for vineyard composting, which returns part of the removed nutrients to the soil for reuse (Harbertson et al., 2008).

Tannins are made up of a large number of subunits connected in chains. Shorter forms of tannin chains can sometimes have a more pronounced bitter taste compared to long tannin chains, which is characteristic of tannins from seeds that have a stronger degree of polymerization. The tannin chains of the tannins of the seeds are shorter and consist of 4–20 subunits, while the tannin chains of the skin are long with 25–100 subunits. With an increase in the length of the chains and the polymerization of tannins, the bitterness and pungency of the wine increase linearly (Cheynier et al., 2006). The astringency of wine can also be expressed by longer ageing, aging of wine as well as an increase in total acid content (Brossaud et al., 2001). With the increase in the length of the tannin chains, a structural change occurs, which changes the stereochemical position of individual parts within the tannin chain, and thus the activity and physical properties of the tannin. This can explain the fact that the tannins of the skin, precisely because of these chemical and physical properties, are very easily connected to the insoluble matrix, which is mainly composed of pectins of the cell wall and glucans.

Tannins, which are an integral part of grapes and wine, belong to the group of so-called condensed tannins. In the formation of tannins, leucoanthocyanidins (leucoanthocyanidin), catechins and gallic acid products play a key role. Leukoanthocyanins and catechins are close in their chemical composition and are flavan derivatives that differ from most phenols as they occur in nature in the form of aglycones, which are polymeric. The most important flavans are catechins, gallocatechins, leucocyanidins and leucodelphinidins. Catechin, epicatechin and gallocatechin were detected in the skin and seeds. In the seeds, they are most abundant in the esterified form with gallic acid. Esterified forms have pronounced pungency (Hanlin & Downey, 2009). Catechin and its polymers are the main flavour carriers of red wines (Figure 58) and are the source of bitterness (Puškaš, 2010). The concentration of these compounds in the skin is minimal, almost in traces, while the recorded concentration in the seeds is disproportionately higher compared to the skin. The reason is that after fertilization and setting of berries under the influence of light, there is inactivation of the gene for tannin synthesis in the skin and activation of the gene for tannin synthesis in the seeds (Downey et al., 2006). With the growing degree of ripening and ripening of the seeds in the berry, the polymerization and oxidation of tannins occur, which is reflected in the appearance of the dark colour of the pistil from the hinge to full maturity. The seed tannins become a strong barrier between the rest of the tannins inside the seed and do not allow the transfer of the entire amount of tannins into the wine. Based on this, it can be concluded that the amount of extracted tannin in the wine and the astringency of the wine are correlated with the number of seeds in the berry. The degree of astringency and pungency of tannins depends on the degree of polymerization of leucoanthocyanins. The tartness and pungency of red wines are also related to the presence of gallic and ellagic acids. Of all the catechins, D-catechin takes the main place. The gallic acid in the composition of tannins is most often bound in the ester form and its most important representative is D-catechin gallate. D-catechin gallate in most varieties makes up 1.5–5% of the total amount of tannins.

The importance of tannic substances is multiple, but the physiological importance of tannins can be emphasized above all. Tannins are subject to the oxidation process and as such participate in the breathing process of the berry. In a certain concentration, they have a favourable effect on the taste of grapes, grape juice and wine (Figure 59).

Some tannins are attributed to the properties of vitamins (vitamin P), with leucocyanidin and some isomers of catechins being particularly prominent. The importance of tannic substances is also reflected in some of the physicochemical and biochemical processes of wine stabilization, where they directly affect the degree of protein deposition. Tannins are also of great importance in the biological stabilization of wine, where they inhibit the work of certain representatives of the wider wine flora. During longer storage of grapes, the tannins from the seeds move into the mesocarp of the berries and thus the berries become astringent. For this reason, it is recommended that seedless grapevine varieties are more suitable for storage (Ribereau-Gayon et al., 2006). Biosynthesis of tannins is performed immediately after fertilization and setting of the berries, i.e., during the hottest part of the vegetation, and with the appearance of the hinge, there is an interruption in the synthesis of tannins and their polymerization. Along with the cessation of tannin synthesis, the synthesis and concentration of anthocyanin in the skin of the berry begins (Hanlin & Downey, 2009).

According to Souquet et al., (2000), tannins are contained in all organs of the grapevine, with the richest being those organs in which the exchange of matter is most dynamic. The highest amounts of tannins were recorded in flowers, leaves, buds, and roots. The tannin content also changes according to the phenophases of the grapevine's development. During the phenophase of the development of green berries, the concentration of tannins is the highest and then decreases with ripening. Observing the proportion of tannins and their distribution in the berry, it can be concluded that they are most abundant in the pulp, seeds, skin and finally in the mesocarp (mesocarp 5–10%, skin 26–30%, seeds 45–57%, skin 43–50%). The tannin content varies by variety. The concentration of tannins in wine depends on the way the grapes are processed, the length of maceration, as well as whether or not the bunches are separated from the stem.

4. Coloured Substances in Grapes

Coloured substances originate from plant pigments chlorophyll and carotenoids (located in plastidial cells) but also anthocyanins and flavone pigments (dissolved in cell juice). Anthocyanins can also be located in special intramembrane structures of anthocyanoplasts where they are protected from oxidation and degradation. Most varieties of grapevines have pigments located in the skin, except the varieties of Bojadiser (*Game bojadiser, Alicante bush*) in which pigments are also found in the mesocarp of the berry. Chlorophyll is a pigment that is present in all varieties of grapevines until the hinge when it disappears and remains present only in the grape skin until before the grapes are fully ripe when it breaks down and lignin appears, which causes the dark colour of the skin. In certain table varieties, the decomposition of chlorophyll is followed by the lignification of the petiole (*Afus ali*). Until the synthesis of the other coloured pigments and their clear perceptibility, it starts with the hinge so that their concentration reaches the maximum when the grapes are fully ripe. All varieties are divided into white and black based on the presence of coloured pigments. Depending on the type of coloured pigments, the skin can have different shades. Thus, the following shades are found in white varieties: green-yellow, yellow-green, whitish-yellow, greyish-yellow, light yellow and dark yellow. In black varieties, the skin can be: light blue, grey-blue, reddish-blue, violet-blue, dark blue, light pink, pink, red, carmine red, purple-red, yellow-red, etc. (Castillo-Munoz et al., 2007).

4.1. Coloured Substances of Black Grapes and Red Wines

The various shades of coloured pigments of red grapes that appear from the hinges come from anthocyanins, flavonoid compounds that rank second in importance to tannins. The synthesis and accumulation of anthocyanins begin with the appearance of the verasion, where the main signal for the start of anthocyanin synthesis is the beginning of the accumulation of sugar in the berry (Castellarin et al., 2007b).

Grape anthocyanins originate from three basic anthocyanidins that differ from each other according to the number of OH groups in the lateral phenyl ring: pelargonidin contains one OH group, cyanidin-contains 2 OH groups and delphinidin-contains 3 OH groups. Pelargonidin is a red anthocyanin, cyanidin is dark red, and delphinidin is pinkish purple. Derivatives of these three anthocyanins are formed by methylation of hydroxyl groups. In grapes, delphinidin glucosides are most common, while cyanidin glucosides and glucosides of their derivatives: peonidin, petunidin and malvidin are much less common. Glucoside malvidin is the main ingredient of most black grape varieties (Pomar et al., 2005). Black grapes of European varieties contain the most common monoglucosides delphinidin, malvidin, petunidin, cyanidin and peonidin. The red colour of young red wines is given by the phenolic compounds anthocyanins found in the skin of black grapes. Only a small number of varieties (so-called dyers) contain anthocyanins in the juice, that is, they have a red-coloured juice. During maceration, which is a regular procedure in the production of red wines, the anthocyanins from the skin of the grain pass into the juice, that is, the wine. Although the primary purpose of maceration is the separation of coloured substances from the skin of the berries, together with them from the skin, but also the seeds and other substances are also separated.

Thus, red wines obtained with long or intensive macerations have higher amounts of tannins, minerals, often acids, etc. (Figure 60). The fact that the red-coloured substances in most black grape varieties are found only in the skin, allows the production of white wine from black grapes. If the grapes are quickly crushed and squeezed, a partially colourless must will be obtained, which after alcoholic fermentation will become a type of white wine. Practice, as well as many researchers, indicate a significant reduction in the colour of red wines already during alcoholic fermentation. As red wines mature and age, anthocyanins bind to other phenolic compounds forming red-coloured polymers (large complex molecules) that, among other things, are more resistant to oxidative discolouration or fading of the wine under the influence of sulfur dioxide.

Anthocyanins along with tannins may also be included in high molecular weight polymers. As the size of the polymer grows, the colour of the wine changes from intense red or purple-red to brown-red. It is considered that free anthocyanins are not the carriers of the colour of old red wines at all, but that they are mostly tannins or polymers of tannins and anthocyanins (Figure 61).

The various shades of coloured pigments of red grapes that appear from the verasion come from anthocyanins, flavonoid compounds that rank second in importance to tannins. The term "anthocyanin" comes from the Greek words *anthos* - flower and *kyaneos* - blue. The synthesis and accumulation of anthocyanins begin with the appearance of the verasion, where the main signal for the beginning of anthocyanin synthesis is the beginning of the accumulation of sugar in the berry (Castellarin et al., 2007b). Anthocyanins are found in grapes mainly in the form of glucosides-anthocyanosides, which contain one or two sugar molecules and distinguish between mono (3-glucosides) and diglucosides (3,5-diglucosides). A non-sugar aglycone component called anthocyanidin is found as an integral part (Bowles et al., 2006). The basic core of anthocyanidin is benzopyrillium chloride, it is the original compound from which 2-phenylbenzopyrillium chloride (flavium chloride) is formed. In that form, they are chemically more stable and soluble in water, and they can also be bound with acetic, coumarin and caffeic acid. All anthocyanidins are derivatives of trihydroxyflavylium chloride, and they have in common that they have two rings in the molecule: a basic benzopyrylium ring or core and a phenyl ring as a side ring. Functional groups are located around the nucleus: hydroxyl (-OH) or methoxy group ($-OCH_3$). The differences are due to substitution and the different number of groups on the 2-phenyl ring.

The number and position of hydroxyl or methoxy groups affect the colour and stability of anthocyanins. If the hydroxyl group is in the R position, the anthocyanins will be coloured red, but if the hydroxyl group is in the R and R` positions, the anthocyanins will be blue coloured. The enzyme flavonoid hydroxylase plays a key role in the position of hydroxyl groups, and therefore the degree of anthocyanin colour, whose synthesis is genetically determined (Tanaka et al., 2008). Grape anthocyanins originate from three basic anthocyanidins that differ from each other according to the number of OH groups in the lateral phenyl ring: pelargonidin-contains one OH group, cyanidin-contains 2 OH groups and delphinidin-contains 3 OH groups. Pelargonidin is a red anthocyanin, cyanidin is dark red, and delphinidin is pinkish purple. Derivatives of these three anthocyanins are formed by methylation of hydroxyl groups. In grapes, delphinidin glucosides are most common, while cyanidin glucosides and glucosides of their derivatives: peonidin, petunidin and malvidin are much less common. Glucoside malvidin is the main ingredient of most black grape varieties.

The colour largely depends on the structure of the glucoside, which changes with pH, so in a more acidic environment, a red colour stands out, while at higher pH shades of blue-violet colour come to the fore. Also, the shade of colour depends on the degree of hydroxylation of the B-ring, as well as the ratio of free and methylated hydroxyl groups, and the presence of SO_2 can lead to bleaching.

According to Castellarin & Di Gaspero (2007c), based on the diverse content and different concentrations of certain mentioned pigments, we can talk about the so-called "anthocyanin complex." Black grapes of European varieties contain the most common monoglucosides delphinidin, malvidin, petunidin, cyanidin and peonidin. Varieties are rich in Malvidin: *Shiraz* and *Burgundac black* as well as most black table grapes. Cyanidin is dominant in the skin of the *Pinotage* variety, peonidin in the *Nebiolo* variety, and cyanidin and delphinidin in the skin of the *Concord* variety. *Merlot, Moorweather* and *Sangiovese* have a uniform ratio of the mentioned pigments in the skin of the berry. Jackson (2008) examined a large number of cultivars and compared them based on the content of total and individual anthocyanins in the skin of the berry. Tests showed wide variations both in total phenolic values and in the degree of anthocyanin accumulation. *Cabernet Sauvignon* recorded the highest anthocyanin values (2,339 mg/kg berries), followed by *Shiraz* (2,200 mg/kg berries), *Malbec* (1,710 mg/kg berries), *Carignan* (1,638 mg/kg berries), *Tempranillo* (1,493 mg/kg berries), *Grenache* (1,222 mg/kg berries), while significantly lower concentrations of anthocyanins were recorded in the varieties *Gama* (844 mg/kg berries) and *Cinsaut* (575 mg/kg berries). *Bojadiseri* varieties are characterized by significantly higher concentrations of anthocyanins compared to the mentioned varieties. *Alicante bush* can reach an anthocyanin concentration of up to 4893 mg/kg of berries. The highest percentage share of individual anthocyanins was recorded for malvidin and peonidin (43–77%), petunidin (3–16%) and delphinidin (4–25%).

American types of grapevines generally contain diglucosides of the mentioned aglycones in grapes, as well as wine, which is a specific oenological indicator that indicates differences in the chemical composition of wine between European varieties and American types. Hybrid varieties like *Concord* contain mono and diglucosides. An exception to these rules is the direct-bred hybrid *Siebel 5455*, which has the same anthocyanin profile as European grape varieties. The intensity of the colour of the berries, i.e., the concentration of coloured substances is in direct correlation with the conditions of the external environment as well as the applied lighting technique. Depending on the mentioned conditions, the colour of berries and wine varies from year to year. The synthesis of coloured substances is also influenced by the variety. Under the same external conditions, certain varieties have pink skin, some red, and some bluish skin. Light plays a key role in anthocyanin synthesis. This is confirmed by the fact that wines from more southern wine-growing areas are more coloured than wines from northern wine-growing areas. The ecological conditions of the locality affect both the composition of the anthocyanin complex and the quantitative ratio of individual groups of anthocyanins (Marković, 2012).

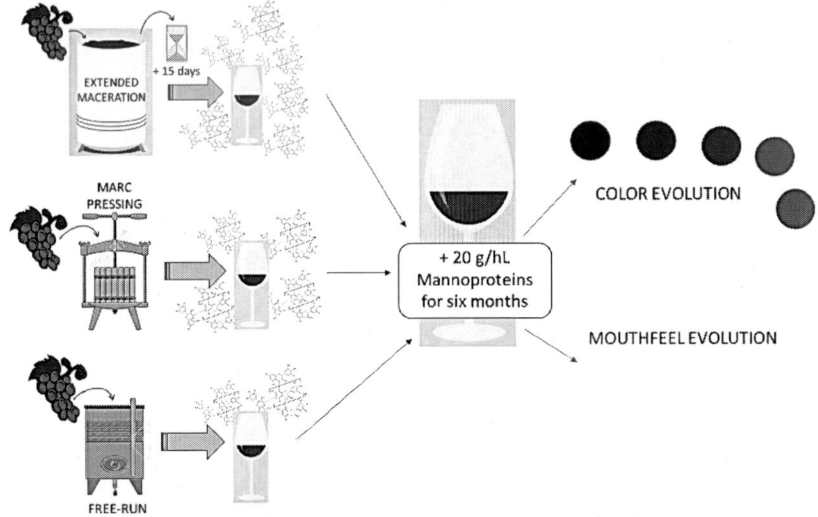

Figure 60. Colour of red wines (Rinaldi et al., 2021).

Anthocyanins are resistant to degradation in direct sunlight. It was found that black varieties with red, purple, blue and black shades of anthocyanins absorb light from the visible part of the spectrum, especially the ultraviolet part, to a greater extent (Cadle Davidson & Owens, 2008). According to Ortega-Regules et al., (2006b) varieties with more acids and higher acidity give open red wines, while varieties with less acid give closed red wines. Anthocyanins dissolve poorly in wine, while with the rise in temperature and the appearance of alcohol during fermentation, they dissolve and pass into the wine. By quickly separating the wider black grapes from the stem, a rose or white wine can be produced. The degree of colouration of the wine depends on the length of maceration. It is believed that 50–90% of the total amount of anthocyanin is extracted into the wine. The degree of extraction also depends on the degree of ripeness of the berry. White wine cannot be produced from dyeser cultivars as their mesocarp is coloured and thus the juice (Fulcrand et al., 2006).

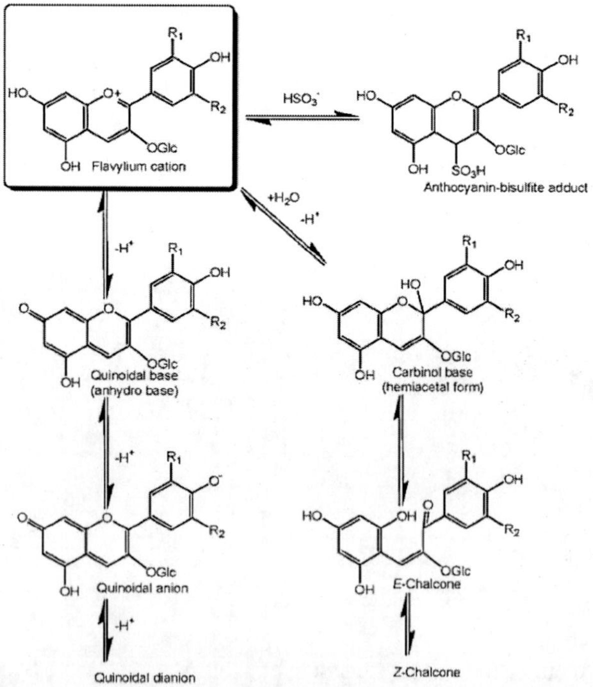

Figure 61. Anthocyanin equilibrium forms found in red wine (Kennedy et al., 2006).

4.2. Coloured Substances on White Grapes

The colour of white grape varieties comes from compounds of a flavonoid character, the synthesis of which begins with the setting of the berry and continues throughout the ripening period. The word "flavus" is of Latin origin and means yellow. In the literature, you can also find the old term for yellow pigments "anthoxanthin," which comes from the Greek words *anthos* - flower and *xanthos* - yellow. Flavonols are concentrated to a greater degree in bunches that are exposed to the sun, so unsunned grapes have a lower content of flavonols in the skin of the berry. A lower concentration of flavonols was also found in the less sunny part of the berry compared to the sunny side of the same berry (Lenk et al., 2007).

Yellow pigments of water originate from flavone glucosides. The skin of white grape varieties contains four yellow pigments - glucosides of the following aglycones: quercetin, myricetin and kaempferol and the glucoronoside of quercetin - quercituron. Quercetin and quercituron are yellow, kaempferol is yellow-green, and myricetin is golden yellow. Quercetin is the dominant flavanol in the skin of white sweet potatoes, myricetin can be specifically dominant only in certain varieties, while syringetin can be found in traces (Downey & Rochfort, 2008).

5. Aromatic Compounds

Aroma represents the most complex chemical component of grapes and wine, especially since it consists of a large number of compounds whose concentration varies under the influence of variety (Prosen et al., 2007a), soil and climate parameters, i.e., in a word, terroir (Figure 62 and Figure 63). In addition to these elements, the grape aroma can be influenced by the application of certain ampelotechnical measures, the application of chemical preparations in the vineyard, the time of harvest, stressful growing conditions such as drought (Sanches-Palomo et al., 2007), the strain of yeast used during fermentation, the type of wood from which the vessel in which the wine is aged, the length of wine ageing (Chalier et al., 2007; Košmerl et al., 2008), etc.

The aromatic compounds of grapes, must and wine comprise many different compounds. Most of these compounds are found in such low concentrations that appropriate analytical equipment is required to detect them. However, a significant number of aromatic compounds whose quantity in wines is measured in nanograms can have a greater influence on the aroma of the wine compared to other aromatic compounds present in significantly higher concentrations.

According to origin, aromatic compounds from grapes are divided into:

- primary aromatic compounds (original or grape aroma) – The primary aroma comes from monoterpenes and C_{13}-norisoprenoids, while in some varieties it is characterized by compounds from the pyrazine group. The synthesis of these compounds begins with the initial accumulation of sugar in the berry (with a verasion), and their concentration increases as the grapes ripen. They are created in the bunch and pass into the wine during processing and maceration and participate in the formation of its particularly pleasant (*Muscat* variety) or unpleasant (hybrid variety) aroma. They are found primarily in the hard parts of the grain, especially in the skin. They are strong aromatic compounds that give the aroma of *Muscat* varieties and the hybrid aroma of grapes of American grapevines and direct native hybrids;
- secondary aromatic compounds – are created during primary processing (grinding of grapes, activity of enzymes). These include substances whose aromas are not particularly expressed in grapes and must. These compounds, as precursors of the aromatic compounds, are transformed during alcoholic fermentation or later during the ageing of the wine, thus obtaining complex specific aromas;
- aroma that is created during fermentation;
- tertiary aromatic compounds – arise in the ageing process of the wine and are called bouquet substances.

In the group of aromatic compounds in wine, special attention is paid to substances that are carriers of the so-called varietal aromatic wines. The amounts and structure of these substances, above all, depend on the grape variety and the conditions in which it was grown and ripened. *Muscat* grape varieties contain specific muscat aromatic compounds that can be felt in the must obtained from such grapes. However, most grape varieties in the must contain only precursors (substances that are not aromatic themselves, but later develop aromatic compounds), i.e., aromatic compounds that are the carriers of the varietal aroma of the wine. This condition is found in almost all the so-called large wine varieties (*merlot, cabernet sauvignon, cabernet franc, chardonnay, pinot*, etc.).

The term varietal aroma does not mean that each grape variety has distinctly different odorous substances. The same aromatic substances or precursors of aromatic compounds are found in the must of a large number of grape varieties, and the specific varietal aroma is a consequence of their different ratios in the wines.

Aromatic compounds in grapes are mostly found in the skin of the berries and the flesh zone immediately below it (Schwab et al., 2008), so in the production of both white and red wine, more and more attention is paid to the extraction of these substances from the skin. The ripeness of the grape, its state of health and varietal affiliation largely determine the amounts and structure of aromatic compounds. Good conditions for ripening grapes, especially on some types of soil, result in a harvest with great potential for obtaining wines with well-developed aromas.

Of all the nutrients that are used in the nutrition of the grapevine, potassium and phosphorus have the most favourable effect on the taste and aroma of the grapes.

		Terroir factor			
	Aroma compound or family	Air temperature	Radiation	Vine nitrogen status	Vine water deficit
Green and peppery flavours	IBMP	decrease	decrease	no effect	decrease
	(-)-rotundone	decrease	increase	not yet investigated	decrease
	1,8-cineole	decrease	not yet investigated	not yet investigated but likely to increase	decrese
Other monoterpenes		variable effect	increase	decrease	variable effect
Volatile thiols and C_{13}-norisoprenoids	Volatile thiols	decrease	increase	increase	increase when moderate
	TDN	increase	increase	decrease	increase
	Tabanones	increase	not yet investigated	not yet investigated	increase
	Other C_{13}-norisoprenoids	no effect	increase	increase	increase
Dried fruit aromas		increase	increase	not yet investigated	possible increase through dehydration
Esters		not yet investigated	increase	increase	increase
Other compounds	DMS	increase	not yet investigated	increase	increase
	Red wine aging bouquet	likely to increase	not yet investigated	not yet investigated	increase
	o-aminoacetophenone	not yet investigated	increase	decrease	increase
	Glutatione	increase	decrease	increase	decrease
	Tanins	no consistent effect reported	increase	decrease	increase

Figure 62. Effect of four terroir factors (air temperature, radiation, grapevine nitrogen status, and grapevine water deficit) on aroma compounds in grapes and wines (Leeuwen et al., 2023).

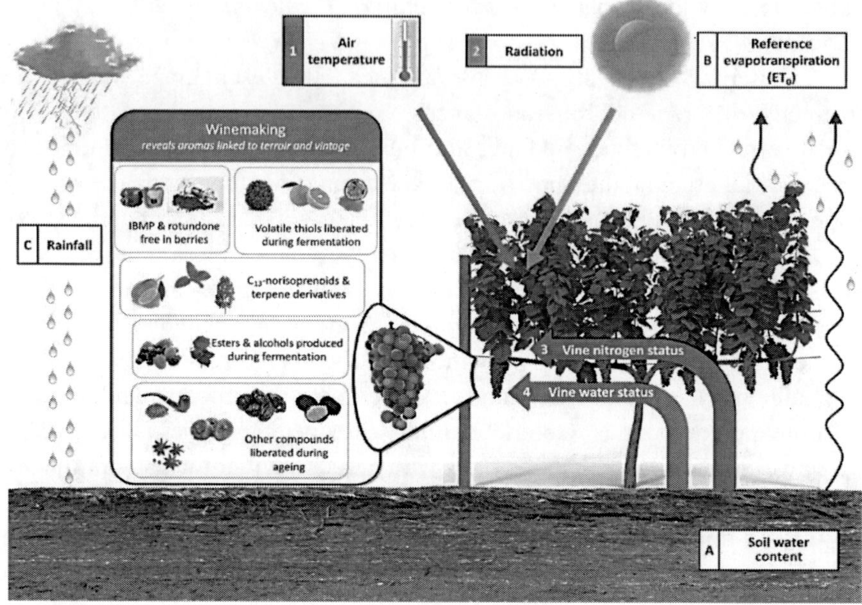

Figure 63. Terroir expression is mainly mediated through (1) air temperature (climate), (2) radiation (climate), (3) grapevine nitrogen status (soil), and (4) grapevine water status, which results from (A) soil water content (soil), (B) reference evapotranspiration (climate), and (C) rainfall (climate). These four components related to soil and climate impact aroma composition and expression in grapes and wines (Leeuwen et al., 2023).

6. Phenolic Compounds in Wine and the Biological Potential of Wine

Numerous studies have shown that wine is the first on the list of alcoholic beverages when it comes to the positive impact of alcoholic beverages on human health. Namely, it has been proven that wine has a stronger antibacterial effect against the Gram-negative bacteria *Vibrio cholera* and *Escherichia coli* than ethanol and gin, while compared to ethanol, beer, whiskey and other strong drinks, in animal models of atherosclerosis, wine showed the best effect on inhibition of disease development. It was also investigated how much alcohol contributes to the activity of wine, where by comparing wine, de-alcoholized wine and grape juice, it was concluded that the specific chemical composition of wine, and not ethanol, is responsible for its beneficial effect on human health (Vinson et al., 2001; Cordova et al., 2005; Stojanova et al., 2024).

Since wine is an alcoholic beverage, attention has also been focused on other products of the grapevine such as grapes, grape juice and grape seed extracts, due to their similar chemical composition to wine, without the side effects that alcohol can cause. It has been proven that these products can also be effective in the treatment of cardiovascular and neurodegenerative diseases, cancer, hypertension and hyperlipidemia. In addition, many researchers believe that wine has a better effect in the mentioned conditions, precisely because of the presence of alcohol, which contributes to better extraction of phytonutrients during fermentation, as well as their increased solubility and bioavailability in the body. In epidemiological studies related to the protective effects of alcoholic beverages on human health, a curvature has been observed representing the relationship between alcohol intake and mortality, indicating that individuals who consume alcoholic beverages in moderation have a lower risk of developing the aforementioned diseases than those who abstain or they consume too much of them. Therefore, moderate consumption of wine is advocated, which means one glass of wine (≈150 mL) for women, and two glasses for men (≈300 mL), with the recommendation that the intake be before or with the evening meal, for the most pronounced effect (O'Keefe et al., 2014).

Red wines are obtained from grapes that contain substances of red colour - anthocyanins. In most varieties, these substances are found only in the skin of the grape berry. There is a small number of economically significant varieties in which, in addition to the skin, anthocyanins are also found in grape juice. In the process of making red wines, the lees are most often subjected to alcoholic fermentation, and the wine is separated from the overboiled lees by the processes of draining and straining. An exception to the above is the technologies according to which coloured substances and other ingredients are extracted from the solid parts of the stem by special procedures, and in the further course, the wines are produced according to the technological scheme for the production of white wines (Blesić, 2001).

In the technological process of red wine production, during maceration and extraction, dyes and other substances pass from the pomace into the liquid fraction, which is the basis of the differences in the quality elements of white and red wines (Figure 64). Other technological operations in the production of white and red wines do not differ in essence. Flavonoids, i.e., anthocyanins and proanthocyanidins make up the largest part of red wine polyphenols that affect the sensory quality of red wines. They are wine preservatives and the basis for ageing. Anthocyanins are responsible for the red colour, and proanthocyanidins for colour stability, bitter taste and astringency in the mouth. In general, the astringency of proanthocyanidins increases with chain length, and the bitterness decreases. Other studies, however, report that the differences in astringency are mainly due to the total proanthocyanidin content of the wines. Flavonoids, besides having technological importance, are also powerful antioxidants that are important for human health (Rodrigo et al., 2011). The levels of anthocyanins and proanthocyanidins and the distribution of proanthocyanidins between the skin and seeds of grape berries are determined according to the grape variety and according to climatic and pedological conditions (Mattivi et al., 2009).

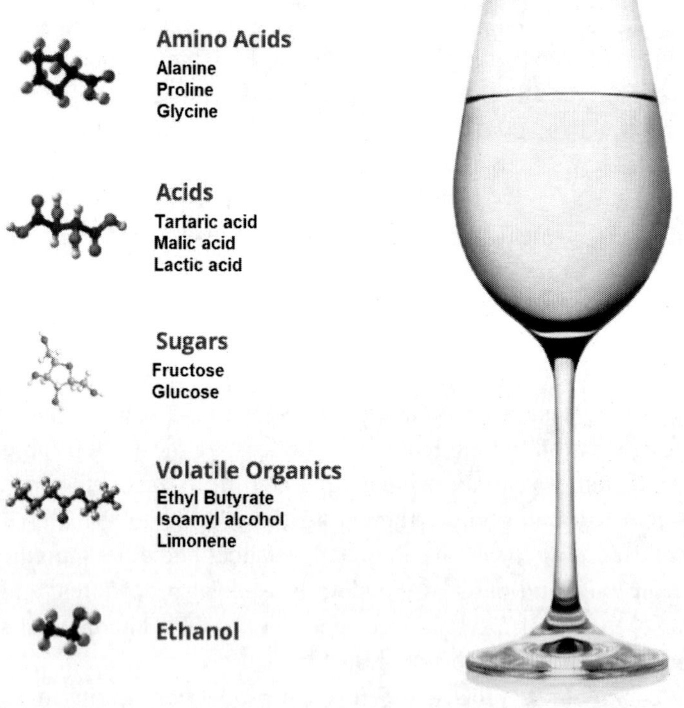

Figure 64. Chemical components of red wines.

Both anthocyanins and tannins simultaneously participate in the creation of the colour of young red wines, with the more pronounced influence of anthocyanins. Due to wine storage, there is a constant decrease in anthocyanin concentration. The main flavour carriers of red wines are catechin and its polymers, while catechin monomers are sources of bitterness and astringency. Proanthocyanidols give a feeling of roughness and dryness in the mouth. Tannins have a significantly greater influence on the taste of red wines than anthocyanins. The solubility of tannins in wines is increased by polymerization with anthocyanins. The fullness of taste depends on the content of phenolic compounds, and they indirectly affect the experience of sweet and sour tastes. The balanced taste of red wines is very important. The sweet taste, which originates from alcohol, polysaccharides and sugar, must be balanced with acidity, bitterness and astringency. Bitterness, pungency and astringency of wine are always observed together with the taste of other components of wine (Puškaš, 2010). In the skin of black and white grapes, the most abundant class of flavonoids are flavonols. They are also the strongest antioxidants in wine, especially white wine, however, flavan-3-ols (catechins) and anthocyanins are more prevalent in red wine. The colour of white wine is contributed by flavonols, which are mostly yellow pigments, but red pigments - anthocyanins - are more dominant in red wine.

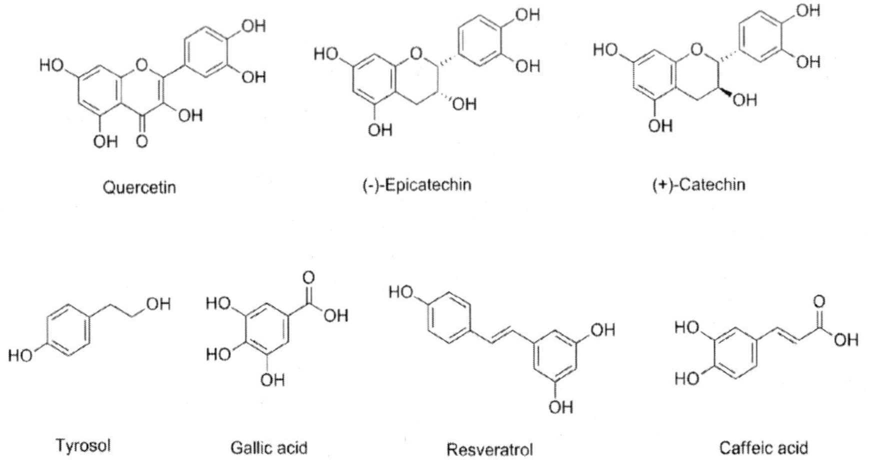

Quercetin

(-)-Epicatechin

(+)-Catechin

Tyrosol

Gallic acid

Resveratrol

Caffeic acid

Figure 65. Examples of wine phenolic compounds with biological activities (Martínez-Huélamo et al., 2017).

The importance of flavonols in wine production is that, together with anthocyanins, they participate in the copigmentation process and support their extraction during vinification, leading to an increase in the intensity of the wine's purple colour (Figure 65). Spayd et al., (2002) state that the content of flavonols increases several times in plants intensively exposed to sunlight. The final content of flavonols can be influenced by the applied agrotechnical measures during grape production and wine production procedures. They are easily degraded if exposed to heat, enzymes and oxidants because they are less stable molecules in their free form (Đorđević, 2020). Castellari et al., (2000) state in their research that the level of quercetin during wine storage decreases by more than 50% in six months with oxygen supplementation, and that the level of quercetin decreases significantly in wine samples stored at 22°C, compared to those stored at 12°C.

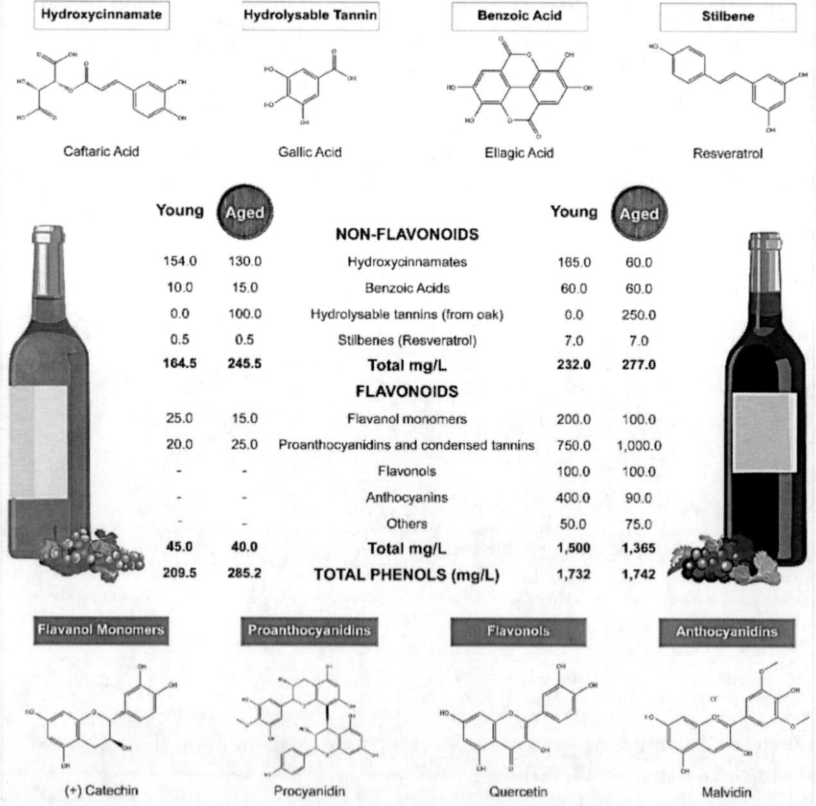

Young	Aged		Young	Aged
		NON-FLAVONOIDS		
154.0	130.0	Hydroxycinnamates	165.0	60.0
10.0	15.0	Benzoic Acids	60.0	60.0
0.0	100.0	Hydrolysable tannins (from oak)	0.0	250.0
0.5	0.5	Stilbenes (Resveratrol)	7.0	7.0
164.5	**245.5**	**Total mg/L**	**232.0**	**277.0**
		FLAVONOIDS		
25.0	15.0	Flavanol monomers	200.0	100.0
20.0	25.0	Proanthocyanidins and condensed tannins	750.0	1,000.0
-	-	Flavonols	100.0	100.0
-	-	Anthocyanins	400.0	90.0
-	-	Others	50.0	75.0
45.0	**40.0**	**Total mg/L**	**1,500**	**1,365**
209.5	**285.2**	**TOTAL PHENOLS (mg/L)**	**1,732**	**1,742**

Figure 66. Red and white wine phenolic composition (Silva & Vauzour, 2018).

Quercetin is also the most common flavonol in grapes and its content in black grapes can be up to 87% of the total flavonol content (Mattivi et al., 2006). Myricetin is present only in red wines, while kaempferol and quercetin are also present in white wines. In grapes, anthocyanins are present in the skin and to a lesser extent in the mesocarp, and they are also present in the leaves, at the end of the growing season. Five anthocyanin molecules (cyanidin, peonidin, delphinidin, petunidin, and malvidin) were found in grapes and wine. The most dominant anthocyanin in all grape varieties is malvidin and its concentration is 50–90%. The greatest importance of malvidin is that it participates in the formation of the base colour of red grapes, and then red wine. Ribereau-Gayon et al., (2006) pointed out that grapes and wine contain seven benzoic acids: gallic, vanillic, p-hydroxybenzoic, gentisic, protocatechuic, salicylic and syringic, while traces of salicylic and gentisic acids are present. Gallic acid as a derivative of hydroxybenzoic acid is included in the composition of ellagitannin and gallotannin, and p-hydroxybenzoic acid, vanillic acid and syringic acid are integral parts of lignin. Depending on the variety and agroecological conditions, the amount of hydroxybenzoic acid derivatives in wines varies, and gallic acid is the most common among them.

Hydroxycinnamic acids are also often present, which are easily extracted from the skin and pulp, and are mostly in the form of esters with tartaric acid, but they can also be bound with other organic acids, alcohols and sugars. Hydroxycinnamic acids are important compounds in the oxidative processes of wine during which anthocyanins and flavan-3-ols are formed, they are very important and the colour of red wines depends on their presence. Hydroxycinnamic acids participate with anthocyanins in the copigmentation process, as a result of which the purple colour of the wine intensifies (Boulton, 2001).

Grapes usually contain camphor acid, which is subject to oxidation and contributes to the darkening of the juice, but in small amounts, oxidized camphor and cutaric acid give the wine a golden yellow colour. This shade can also be the result of the melanoidin's presence, and if the wines have been aged in a wooden barrel, the extracted hydroxybenzoic and hydroxycinnamic acids, as well as lignins, can affect the colour of the wine. Thus, ellagic acid and gallotannins from oak can improve the red colour of the wine by co-pigmentation with anthocyanins, while the degradation of lignin releases various volatile components. The coumarin glucosides esculin and scopolin can also be extracted, which turn into aglycones as the wine ages and thereby reduce the bitterness of the wine that has been aged in a wooden barrel. Some compounds are characteristic of certain grape varieties, such as methyl anthranilate, which gives *V. labrusca* grapes the name fox grapes, and 2-phenyl ethanol, which gives *V. rotundifolia* a rose-like scent. The most important aromatic components are esters of hydroxycinnamic acids, which are formed during fermentation and partly pass into the wine by ageing in a wooden barrel and can also be formed by the metabolism of certain microorganisms. Sometimes eugenol and guaiacol are present with sweet notes of cloves and smoke. Volatile phenolic (Figure 66) acids and aldehydes such as benzaldehyde, vanillin and syringaldehyde are produced by the degradation of lignin from the wooden barrel and are characterized by aromas of almonds, nuts and vanilla (Stojanova et al., 2024).

Monomeric catechins and anthocyanins, proanthocyanidins and some phenolic acids have the highest influence on the antioxidant potential of red wines (Luiz, 2011). Positive correlations were established between the amount of total phenolic compounds and content of gallic acid, (-)-epicatechin, (+)-catechin and antioxidant potential of wine, and the proportion of anthocyanins and antioxidant activity of some red wines (Jackson, 2008). The same wine varieties have a higher antioxidant potential in older wines, which is explained by the difference in the concentration of tannins, which are created as a result of wine maturation. A balanced diet reduces the risk of numerous health problems and is therefore of great importance for the overall health of people. A growing body of data supports the view that the use of medicinal foods, including red wine, can have a beneficial effect on the functioning of the human body. Red wines are one of the most important sources of dietary polyphenolic antioxidants, compounds with proven effects against diseases associated with oxidative stress. Among alcoholic beverages, red wine has been reported to have the highest protective effect against oxidative stress (Woraratphoka et al., 2007).

It was found that its consumption (together with olive oil) is one of the key explanations for solving the "French paradox," i.e., low incidence of cardiovascular disease despite the intake of fatty foods. Polyphenols such as flavonoids, anthocyanins, hydroxybenzoate, hydroxycinnamates and stilbenes, as the main compounds of the wine, are well known for their role in reducing the risk of diseases related to oxidative stress such as cardiovascular, neurodegenerative and cancer. Moreover, these compounds also show good anti-inflammatory properties. Not only do phenolic compounds have antioxidant activity, but also some of their metabolites are antioxidants.

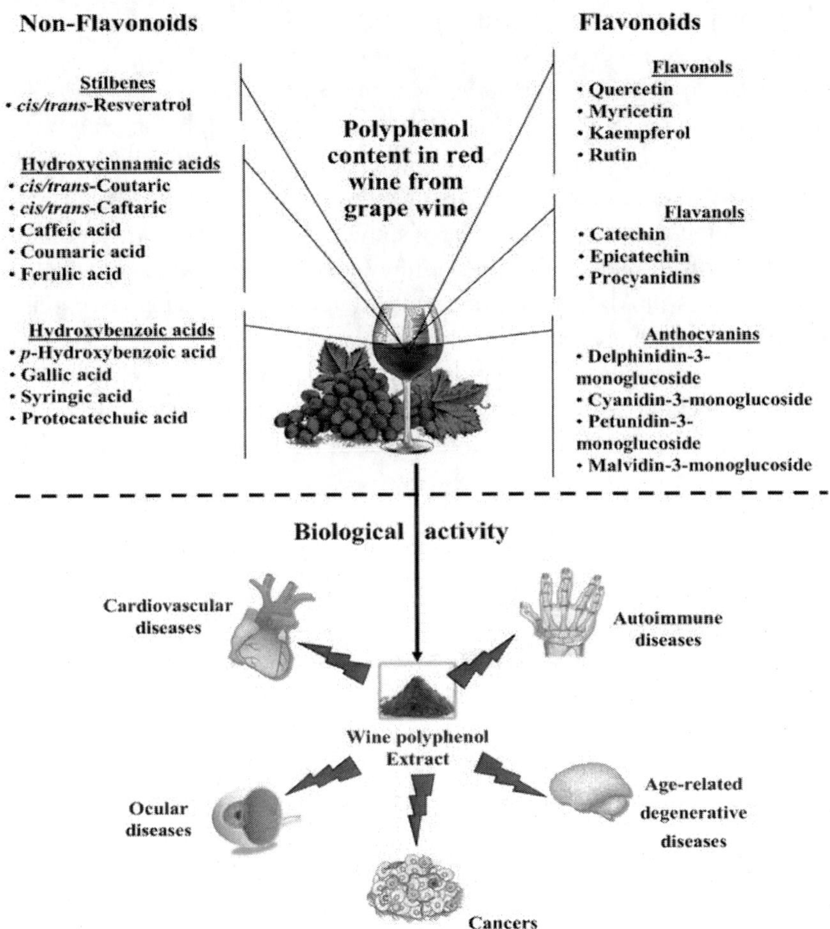

Figure 67. Major constituents in red wine from grapes and the potential biological effects against various diseases (Amor et al., 2018).

The beneficial effects of grapes, juice and wine on human health are most often linked to the presence of polyphenolic compounds. These secondary metabolites play a significant role in the functioning of plants and their protection against pathogens, UV radiation and abiotic stress in the form of unfavourable climate, drought, and excessive exposure to light and pollutants. In addition to strong antioxidant activity, polyphenols in the human body can also act as signalling molecules, to modify the redox status of cells, induce apoptosis and inhibit angiogenesis and proliferation of cancer cells, to influence the inflammatory response through the activation or inhibition of the corresponding enzymes, as well as to modulate genes that are essential for the cell cycle (Figure 67) (Carmo et al., 2018; Stojanova et al., 2024).

Of the compounds found in wine and grape juice, myricetin, morin, quercetin, kaempferol, catechin and epicatechin performed well *in vitro* experiments, which in the study by Ono et al., (2003) showed an influence on the formation, elongation and destabilization of β-amyloid fibrils.

Epigallocatechin gallate has been shown to participate in the regulation of iron metabolism that is disturbed in the brain and leads to the formation of senile plaques in Alzheimer's disease, through a good iron chelating ability, by increasing both protein and mRNA levels of the transferrin receptor, as well as by post-transcriptional reduction of the amyloid precursor protein and toxic β-amyloid peptides. Quercetin has shown a protective effect against neurons from toxicity caused by β-amyloid peptides by modulating oxidative stress and reducing protein oxidation, lipid peroxidation and apoptosis (Ansari et al., 2009).

Another strategy for the treatment of neurodegenerative diseases is to find reverse inhibitors for the enzymes acetylcholine esterase (increased hydrolysis of acetylcholine) and butyrylcholinesterase (similar activity to acetylcholinesterase, increased amounts of this enzyme have been found in senile plaques in Alzheimer's disease). In this respect, quercetin, chlorogenic and gallic acid, kaempferol, hyperoside, quercetin, quercetin-3-O-glucoside, quercitrin, rutin, myricetin and anthocyanins, as well as red wine extracts, also showed in vitro activity. In work with experimental animals, polyphenols from grape seeds inhibited the formation of β-amyloid oligomers and thus slowed down cognitive damage, and it was also observed that they prevent the binding of tau peptides into neurotoxic aggregates and disrupt extracellular signalling in the brain, which slows down the development of tau neuropathology, which characterized by Alzheimer's disease (Wang et al., 2013).

Vepsäläinen et al., (2013) showed in their work that anthocyanins exert their neuroprotective effect through their effect on the metabolism of amyloid precursor protein and β-amyloid peptides, but not tau protein, to alleviate behavioural disorders in mice that represent Alzheimer's disease model systems, and pointed out are the importance of long-term supplementation to achieve the desired health effects. This was also confirmed by Wang et al., (2006) and Ho et al., (2013) in research, where they showed that regular and moderate wine intake in mice prevents the accumulation of β-amyloid peptides and slows the decline of cognitive functions, and found that there is an accumulation of quercetin 3-O-glucuronide in the brain tissue, which prevented the formation of toxic amyloid oligomers and led to improvement of basal synaptic transmission in the hippocampus. The most important observation was that the described positive effects were manifested only in the group that consumed wine, while in the control groups that were given water or ethanol in a concentration equivalent to wine, there was no improvement. In the study of Parkinson's disease, resveratrol showed a significant effect on experimental animals. Intravenous as well as supplemental resveratrol prevented impairment of motor coordination in mice induced by a Parkinson's disease-specific neurotoxin. Also, by exhibiting antioxidant activity, resveratrol protected neurons from hydroxyl radicals, regulated the activity of antioxidant enzymes and reduced the loss of dopamine.

In two clinical studies, one followed the immediate effect of an ingested dose of resveratrol in healthy patients, and in the other the effects of a longer period of supplementation of 52 weeks in patients suffering from mild and moderate forms of Alzheimer's disease, it was proven that resveratrol improves cerebral blood flow, but that it has no effect on the subjects' cognitive functions and biomarkers characteristic of Alzheimer's disease (Turner et al., 2015). In a study examining the effect of daily intake of Concord grape juice (6-9 mL/kg per patient) on patients with a mild decline in cognitive functions (early stage Alzheimer's disease), a moderate improvement in the patient's cognitive functions was seen in verbal and spatial memory tests.

Several epidemiological studies have followed the relationship between the amount and type of alcoholic beverages consumed and the frequency of neurodegenerative diseases. And so, in the Rotterdam study, following a period of 6 years, they saw that moderate consumption of alcoholic beverages (1–3 drinks per day) is proportional to a lower risk of developing any form of dementia, but that there is no relationship between the type of alcoholic beverage and reduced risks. In contrast, two studies involving respondents from Bordeaux and Copenhagen pointed out that regular and moderate wine consumption reduces the risk of dementia (Truelsen et al., 2002).

In the case of neurodegenerative diseases, the beneficial effect of moderate consumption of wine, grape juice and supplements of different parts of grapes is observed, especially in the prevention of the mentioned conditions, but further research is necessary to clarify which components are responsible for the positive effects and by which mechanisms they prevent the onset and progression of the disease. Perhaps the most important step is to shed light on the metabolites of polyphenolic compounds that are formed during digestion, their bioavailability, the possibility of passing through the blood-brain barrier and accumulation in the tissues of the organism so that even better and more thorough clinical studies can be conducted (Caruana et al., 2016).

References

Abo, E., LoAy, S. F., Al-Harbi, A. A., Al-Qahtani, N. A., Allam, S. M., Abdein, H. M. & Abdelgawad, M. M. (2022). A comparison of the effects of several foliar forms of magnesium fertilization on 'superior seedless' (Vitis vinifera L.) in saline soils. *Coatings, 12*(2), 201. doi: 10.3390/coatings12020201.

Add, El-R., Treutter, D., Saleh, M., El-Shammaa, M., Abdel-Hamid, N. & Abou-Rawash, M. (2011). Effect of nitrogen and potassium fertilization on productivity and fruit quality of "Crimson Seedless" Grapes. *Agriculture and Biology Journal of North America 2* (2), 330–40.

Ali, I., Wang, X., Abbas, W. M., Hassan, M. U., Shafique, M., Tareen, M. J., Fiaz, S., Ahmed, W. & Qayyum, A. (2021). Quality responses of table grapes 'flame seedless' as effected by foliarly applied micronutrients. *Horticulturae, 7*(11), 462. doi: 10.3390/horticulturae7110462.

Ali, K., Maltese, F., Choi, Y. H. & Verpoorte, R. (2010). Metabolic constituents of grapevine and grape-derived products. *Phytochemistry Reviews, 9*(3), 357–378.

Amor, S., Châlons, P., Aires, V. & Delmas, D. (2018). Polyphenol Extracts from Red Wine and Grapevine: Potential Effects on Cancers. *Diseases*, *6*(4), 106. https://doi.org/10.3390/diseases6040106.

Anastasiadi, M., Pratsinis, H., Kletsas, D., Skaltsounis, A. L. & Haroutounian, S. A. (2010). Bioactive non-coloured polyphenols content of grapes, wines and vinification by-products: Evaluation of the antioxidant activities of their extracts. *Food Research International 43*, 805–813.

Ansari, M. A., Abdul, H. M., Joshi, G., Opii, W. O. & Butterfield, D. A. (2009). Protective effect of quercetin in primary neurons against Aβ(1–42): relevance to Alzheimer's disease. *The Journal of Nutritional Biochemistry, 20*(4), 269–275.

Arrobas, M., Ferreira, I. Q., Freitas, S., Verdial, J. & Rodrigues, M. A. (2014). Guidelines for fertilizer use in vine yards based on nutrient content of grapevine parts. *Scientia Horticulturae 172,* 191–198. doi: 10.1016/j.scienta. 2014. 04.016.

Azuma, A., Yakushiji, H., Koshita, Y. & Kobayashi, S. (2012). Flavonoid biosynthesis-related genes in grape skin are differentially regulated by temperature and light conditions. *Planta, 236,* 1067–1080. doi: 10.1007/s00425 012-1650-x.

Ball, K. R., Baldock, J. A., Penfold, C., Power, S. A., Woodin, S. J., Smith, P. & Pendall, E. (2020). Soil organic carbon and nitrogen pools are increased by mixed grass and legume cover crops in vineyard agroecosystems: Detecting short-term management effects using infrared spectroscopy. *Geoderma, 379,* 114619. https://doi.org/10.1016/j.geoderma.2020.114619.

Battilani, A. & Mannini, P. (2000). Grapevine (Vitis vinifera) yield and quality response to irrigation. *Acta Horticulturae, 537*, 895–902.

Baydar, N. G. & Akkurt, M. (2001). Oil Content and Oil Quality Properties of Some Grape Seeds. *Turk. J. Agric. For., 25*, 163-168.

Bindon, K., Myburgh P., Oberholster A., Roux K. & Toit C. D. (2011). Response of grape and wine phenolic composition in Vitis vinifera L. cv. Merlot to variation in grapevine water status. *S. Afr. J. Enol. Vitic. 32*, 71-88.

Bindon, K., Dry, P. & Loveys, B. (2008). Influence of partial rootzone drying on the composition and accumulation of anthocyanins in grape berries (Vitis vinifera cv. Cabernet sauvignon). *Australian Journal of Grape and Wine Research, 14*, 91-103.

Blesić, M. (2001). The stability of colored substances and the quality of wine depending on the conditions of the maceration of the Blatina knok. *Doctoral dissertation*. Faculty of Agriculture, University of Sarajevo.

Bogs, J., Jaffé, F. W., Takos, A. M., Walker, A. R. & Robinson, S. P. (2007). The grapevine transcription factor VvMYBPA1 regulates proanthocyanidin synthesis during fruit development. *Plant Physiol. 143*, 1347-1361.

Bojnec, Š. (2006). Wine markets in central Europe. *J. Cent. Eur. Agric., 7,* 465–743.

Bombardelli, E. & Morazzoni, P. (1995). Vitis vinifera L. *Fitoterapia 66*, 291-317.

Boskovic, I., Djukic, D., Maskovic, P. & Mandic, L. (2022). Influence of solvent type on the phenolic content and antimicrobial and antioxidant properties of Echium vulgare extracts. *Facmacia, 70*(4), 665–670. https://doi.org/10.31925/farmacia.2022.4.12.

Boulton, R. (2001). The copigmentation of anthocyanins and its role in the color of red wine: A critical review. *American journal of enology and viticulture, 52*(2), 67-87.

Bowles, D., Lim, E. K., Poppenberger, B. & Vaistij, F. E. (2006). Glycosyltransferases of lipophilic small molecules. *Annual Review of Plant Biology, 57*, 567-597.

Braidot, E., Zancani, M., Petrussa, E., Peresson, C., Bertolini, A., Patui, S., Macrì, F. & Vianello, A. (2008). Transport and accumulation of flavonoids in grapevine (Vitis vinifera L.). *Plant signaling & behavior, 3*(9), 626-632.

Bredun, M. A., Gomes T. M., Assumpc¸ T. I., Brighenti, A. F. Chaves, E. S. Panceri, C. P. & Burin, V. M. (2021). Statement of Boron application impact on yield, composition and structural properties in Merlot grapes. *Scientia Horticulturae 288*, 110364. doi: 10.1016/j.scienta.2021.110364.

Brossaud, F., Cheynier, V. & Noble, A. C. (2001). Bitterness and astringency of grape and wine polyphenols. *Australian Journal of Grape and Wine research, 7*, 33-39.

Brunetto, G., Ceretta, C. A., de Melo, G. W. B., Girotto, E., Ferreira, P. A. A., Lourenzi, C. R., da Rosa Couto R., Tassinaria, A., Knevitz Hammerschmitt R., da Silva, L. O. S., Lazzaretti B. P., de Souza Kulmann M. S. & Carranca, C., (2016). Contribution of nitrogen from urea applied at different rates and times on grapevine nutrition. *Scientia Horticulturae, 207*, 1-6. doi:10.1016/j.scienta.2016.05.002.

Brunetto, G., Ceretta, C. A., Kaminski, J., de Melo, G. W. B., Lourenzi, C. R., Furlanetto, V. & Moraes, A. (2007). Aplicaçao de nitrogênio em videiras na Campanha Gaúcha: produtividade e características químicas do mosto da uva. *Ciência Rural, 37,* 389–393.

Brunetto, G., de Melo, G. W. B., Toselli, M., Quartieri, M. & Tagliavini, M. (2015). The role of mineral nutrition on yields and fruit quality in grapevine, pear and apple. *Revista Brasileira de Fruticultura, 37*(4), 1089–104. doi: 10.1590/0100-2945-103/15.

Brunetto, G., Stefanello, L. O. S., Ceretta, C. A., Couto, R. R., Ferreira, P. A. A., Ambrosini, V. G., Borghezan, M., Comin, J. J., De Melo, G. W., Baldi, E. & Toselli, M. (2018). Nitrogen fertilization of 'Chardonnay' grapevines: Yield, must composition and their relationship with temperature and rainfall. *Acta Horticulturae 1228*, 451–456.

Bruwer, F. A. (2018). *Effect of foliar Nitrogen and Sulphur spraying on white wine composition (Vitis vinifera L. cv. Chenin Blanc and Sauvignon Blanc)*. Department of Viticulture and Oenology, Faculty of AgriSciences.

Cadle-Davidson, M. M. & Owens, C. L. (2008). Genomic amplification of the Gret1 retroelement in white-fruited accessions of wild Vitis and interspecific hybrids. *Theoretical and Aplied Genetics, 116,* 1079-1094.

Calzarano, F., Amalfitano, C., Seghetti, L. & Di Marco, S. (2023). Effect of Different Foliar Fertilizer Applications on Esca Disease of Grapevine: Symptom Expression and Nutrient Content in the Leaf and Composition of the Berry. *Agronomy, 13,* 1355. https://doi.org/10.3390/agronomy13051355.

Canoura, C., Kelly, M. T. & Ojeda, H. (2018). Effect of irrigation and timing and type of nitrogen application on the biochemicalcomposition of Vitis vinifera L. cv. Chardonnay and Syrah gra peberries. *Food Chemistry, 241,* 171–181.

Carmo, M. A. V., Pressete, C. G., Marques, M. J., Granato, D. & Azevedo, L. (2018). Polyphenols as potential antiproliferative agents: scientific trends. *Current Opinion in Food Science, 24,* 26–35.

Caruana, M., Cauchi, R. & Vassallo, N. (2016). Putative Role of Red Wine Polyphenols against Brain Pathology in Alzheimer's and Parkinson's Disease. *Frontiers in Nutrition, 3,* 31.

Castellari, M., Matricardi, L., Arfelli, G., Galassi, S. & Amati, A. (2000). Level of single bioactive phenolics in red wine as a function of the oxygen supplied during storage. *Food Chemistry, 69*(1), 61–67.

Castellarin, S. D. & Di Gaspero, G. (2007c). Transcriptional control of anthocyanin biosynthetic genes in extreme phenotypes for berry pigmentation of naturally occuring grapevines. *BMC Plant Biology, 7,* 46.

Castellarin, S. D., Pfeiffer, A., Sivilotti, P., Degan, M., Peterlunger, E. & Di Gaspero, G. (2007b). Transcriptional regulation of anthocyanin biosynthesis in ripening fruits of grapevine under seasonal water deficit. *Plant, Cell and Environment, 30,* 1381- 1399.

Chalier, P., Angot, B., Delteil, D., Doco, T. & Gunata, Z. (2007). Interactions between aroma compounds and whole mannoprotein isolation from Saccharomyces cerevisae strains. *Food Chem. 100,* 22-30.

Champagnol, F. (1984). Elements de physiologie de la vigne et de viticulture generale. *Rev. Agron. 327,* 124–145.

Cheng, G., Fa, J. Q., Xi, Z. M. & Zhang, Z. W. (2015). Research on the quality of the wine grapes in corridor area of China. *Food Science and Technology (Campinas), 35,* 38-44.

Cheynier, V., Remy S. & Fulcrand H. (2000). Mechanisms of anthoczanin and tannin changes during winemaking and aging. *Proceedings of the American Society of Enology and Viticulture 50th Anniversary Annual Meeting*, Seattle, Washington, 8510-8520.

Cheynier, V., Duenas-Paton, M., Salas, E., Maury, C., Souquet, J. M. & Sarni-Manchado, P. (2006). Structure and properties of wine pigments and tannins. *American Journal of Enology and Viticulture, 57*, 298-305.

Chira, K., Pacella, N., Jourdes, M. & Teissedre, P. L. (2011). Chemical and sensory evaluation of Bordeaux wines (Cabernet-Sauvignon and Merlot) and correlation with wine age. *Food chemistry, 126*(4), 1971-1977.

Cindrić, P., Korać, N. & Ivanišević, D. (2019). *Ampelography and vine selection.* Novi Sad: University of Novi Sad, Faculty of Agriculture.

Cocco, A., Mercenaro, L., Musca, E., Mura, A., Nieddu, G. & Lentini, A. (2021). Multiple effects of nitrogen fertilization on grape vegetative growth, berry quality and pest development in Mediterranean vineyards. *Horticulturae, 7* (12), 530. doi: 10.3390/horticulturae7120530.

Conde, C., Silva, P., Fontes, N., Dias, A. C. P., Tavares, R. M., Sousa, M. J., Agasse, A., Delrot, S. & Geros, H. (2007). *Biochemical changes throughout grape berry development and fruit and wine quality.* https://repositorium.sdum.uminho.pt/handle/1822/6820.

Considine, M. J. & Foyer, C. H. (2015). Metabolic responses to sulfur dioxide in grapevine (Vitis vinifera L.): Photosynthetic tissues and berries. *Frontiers in Plant Science 6* (FEB), 60. doi: 10.3389/FPLS.2015.00060/BIBTEX.

Cordova, A. C., Jackson, L. S. M., Berke-Schlessel, D. W. & Sumpio, B. E. (2005). The cardiovascular protective effect of red wine. *Journal of the American College of Surgeons, 200*(3), 428–439.

Cushnie, Y. P. T. & Lamb, A. J. (2011). Recent advances in understanding of the antibacterial properties of flavonoids. *International Journal of Antimicrobial Agents, 38,* 99–107.

Daccak, D., Lidon, F. C., Pessoa, C. C., Luís, I. C., Coelho, A. R. F., Marques, A. C., Ramalho, J. C., Silva, M. J., Rodrigues, A. P. & Guerra, M. et al., (2022). Enrichment of Grapes with Zinc-Efficiency of Foliar Fertilization with ZnSO4 and ZnO and Implications on Winemaking. *Plants, 11,* 1399. https://doi.org/10.3390/plants 11111399.

Dai, Z. W., Ollat, N., Gomès, E., Decroocq, S., Tandonnet, J. P. & Bordenave, L., et al., (2011). Ecophysiological, genetic, and molecular causes of variation in grape berry weight and composition: a review. *Am. J. Enol. Vitic. 62,* 413–425. doi: 10.5344/ajev.2011.10116.

Dani, C., Oliboni, L. S., Agostini, F., Funchal, C., Serafini, L., Henriques, J. A. & Salvador, M. (2010). Phenolic content of grapevine leaves (Vitis labrusca var. Bordo) and its neuroprotective effect against peroxide damage. *Toxicology in Vitro, 24*(1), 148-153.

Davenport, J. R. & Jones, C. (2016). Comparison of foliar- and soil-applied phosphorus fertilizer in wine grape. *Crops & Soils, 49*(4), 30–32. doi: 10.2134/cs2016-49-0407.

De Pascali, S. A., Coletta, A., Del Coco, L., Basile, T., Gambacorta, G., & Fanizzi, F. P. (2014). Viticultural practice and winemaking effects on metabolic profile of Negroamaro. *Food Chemistry, 161,* 112-119.

Debnath, S., Paul, M., Rahaman, D. M. M., Debnath, T., Zheng, L., Baby, T., Schmidtke, L. M. & Rogiers, S. Y. (2021). Identifying Individual Nutrient Deficiencies of Grapevine Leaves Using Hyperspectral Imaging. *Remote Sens., 13,* 3317. https://doi.org/10.3390/rs13163317.

Delgado, R., Martin, P., del Alamo, M. & Gonzales, M. R. (2004). Changes in the phenolic composition of grape berries during ripening in relation to vineyard nitrogen and potassium fertilisation rates. *Journal of the Science of Food and Agriculture, 84,* 623–630.

Dhaliwal, S. S., Sharma, V., Shukla, A. K., Verma, V., Sandhu, P. S., Behera, S. K., Singh, P., Kaur, J., Singh, H. & Abdel-Hafez, S. H., et al., (2021). Interactive effects of foliar application of zinc, iron and nitrogen on productivity and nutritional quality of Indian mustard (Brassica juncea L.). *Agronomy, 11*, 2333.

Dixon, R. A., Xie D. Y. & Sharma S. B. (2005). Proanthocyanidins – a final frontier in flavonoid research?. *New Phytol. 165,* 9-28.

Djukic, D. & Jemcev, V. T. (2003). *Microbiological biotechnology.* Dereta, Belgrade, Serbia.

Djukic, D., Mandic, L. & Veskovic, S. (2015). *General and industrial microbiology, University of Kragujevac*, Faculty of Agronomy, Cacak, Serbia.

Downey, M. O., Dokoozlian N. K. & Krstic M. P. (2006). Cultural practice and environmental impacts on the flavonoid composition of grapes and wine: A review of recent research. *Am. J. Enol. Vitic., 3*, 257-268.

Downey, M. O. & Rochfort, S. (2008). Simultaneous separation by reversed-phase high performance liquid chromatography and mass spectral identification of anthocyanins and flavonols in Shiraz grape skin. *Journal of Chromatography A, 1201,* 43-47.

Downey, M. O., Harvey, J. S. & Robinson, S. P. (2003). Analysis of tannins in seeds and skins of Shiraz grapes throughout berry development. *Aust. J. Grape Wine Res. 9*, 15-27.

Duan, S., Zhang, C., Song, S., Ma, C., Zhang, C., Xu, W., Bondada, B., Wang, L. & Wang, S. (2022). Understanding calcium functionality by examining growth characteristics and structural aspects in calcium-deficient grapevine. *Scientific Reports, 12*(1), 1–14. doi: 10.1038/s41598-022-06867-4.

Duletic, D. & Mijovic, S., (2014). Yield and quality of grapes of the cardinal variety depending on different foliar fertilizers. *Agriculture and Forestry, 60*(2), 85-91.

Ebeler, S. E. & Thorngate, J. H. (2009). Wine chemistry and flavor: Looking into the crystal glass. *J. Agric. Food Chem. 57*, 8098-8108.

Fernández, K., Vega, M. & Aspé, E. (2015). An enzymatic extraction of proanthocyanidins from País grape seeds and skins. *Food Chemistry, 168*, 7–13.

Fernandez, V., Orera, I., Abadia, J. & Abadia, A. (2009). Foliar iron-fertilisation of fruit trees: Present knowledge and future perspectives – A review. *Journal of Horticultural Science & Biotechnology, 84,* 1–6.

Fernandez, V., Sotiropoulos, T. & Brown, P. (2013). *Foliar fertilization scientific principles and field practices.* International Fertilizer Industry Association: Paris, France.

Ferreira, G. M., Moreira, R. R., Jarek, T. M., Nesi, C. N., Biasi, L. A. & May De Mio, L. L. (2022). Alternative con trol of downy mildew and grapevine leaf spot on Vitis labrusca. *Australasian Plant Pathology, 51*(2), 193–201. doi: 10.1007/s13313-021-00836-7.

Fournand, D., Vicens, A., Sidhoum, L., Souquet, J. M., Moutounet, M. & Cheynier, V. (2006). Accumulation and exctracatbility of grape skin tannins and anthocyanins at different advanced physiological stages. *Journal of Agricultural and Food Chemistry, 54*, 7331-7338.

Fulcrand, H., Duenas, M., Salas, E. & Cheynier, V. (2006). Phenolic reactions during winemaking and aging. *American Journal of Enology and Viticulture, 57*, 289-297.

Gambetta, G. A., Herrera, J. C., Dayer, S., Feng, Q., Hochberg, U. & Castellarin, S. D. (2020). Erratum: The physiology of drought stress in grapevine: The anintegrative definition of drought tolerance. *Journal of Experimental Botany, 71*(18), 5717. doi: 10. 1093/jxb/eraa313.

Gambuti, A., Capuano, R., Lecce, L., Fragasso, M. G. & Moio, L. (2009). Extraction of phenolic compounds from 'Aglianico' And 'Uva di Troia' grape skin and seeds in model solutions: Influence of ethanol and maceration time. *Vitis, 48*(4), 193-200.

Gao, Y., Zietsman, A. J. J., Vivier, M. A. & Moore, J. P. (2019). Deconstructing Wine Grape Cell Walls with Enzymes During Winemaking: New Insights from Glycan Microarray Technology. *Molecules, 24*(1), 165. https://doi.org/10.3390/molecules 24010165.

García-Navarro, F. J., Jiménez-Ballesta, R., Garcia-Pradas, J., Amoros, J. A., Perez de los Reyes, C. & Bravo, S. (2021). Zinc Concentration and Distribution in Vineyard Soils and Grapevine Leaves from Valdepeñas Designation of Origin (Central Spain). *Sustainability, 13,* 7390. https://doi.org/10.3390/su13137390.

Garde-Cerdan, T., Lopez, R., Portu, J., Gonzalez-Arenzana, L., Lopez-Alfaro, I. & Santamaría, P. (2014). Study of the effects of proline, phenylalanine, and urea foliar application to Tempranillo vineyards on grape amino acid content. Comparison with commercial nitrogen fertilisers. *Food Chemistry, 163*, 136–141.

Gautier, A., Cookson, S. J., Hevin, C., Vivin, P., Lauvergeat, V. & Mollier, A. (2018). Phosphorus acquisition efficiency and phosphorus remobilization mediate genotype-specific differences in shoot phosphorus content in grapevine. *Tree Physiology, 38*(11), 1742–1751. doi: 10.1093/TREEPHYS/TPY074.

Gawel, R. (1998). Red wine astringency: a review. *Australian Journal of Grape and Wine Research, 4,* 74-95.

Ghorbani, P., Eshghi, S., Ershadi, A., Shekafandeh, A., & Razzaghi, F. (2019). The possible role of foliar application of manganese sulfate on mitigating adverse effects of water stress in grapevine. *Communications in Soil Science and Plant Analysis, 50(13),* 1550–1562. doi: 10.1080/00103624.2019.1626873.

Gong, P., Zhang, Y. & Liu, H. (2022). Effects of irrigation and N fertilization on N fertilizer utilization by Vitis vinifera L. Cabernet Sauvignon in China. *Water, 14*(8), 1205. doi: 10.3390/w14081205.

González-Barreiro, C., Rial-Otero, R., Cancho-Grande, B. & Simal-Gándara, J. (2013). Wine aroma compounds in grapes: A critical review. *Crit. Rev. Food Sci. Nutr*. doi: 10.1080/10408398.2011.650336.

Guilpart, N., Metay, A. & Gary, C. (2014). Grapevine bud fertility and number of berries per bunch are determined by water and nitrogen stress around flowering in the previous year. *European Journal of Agronomy, 54,* 9-20. doi:10.1016/j.eja.2013.11.002.

Gur, L., Cohen, Y., Frenkel, O., Schweitzer, R., Shlisel, M. & Reuveni, M. (2022). Mixtures of macro and micronutrients control grape powdery mildew and alter berry metabolites. *Plants, 11*(7), 978. doi: 10.3390/plants11070978.

Gutiérrez-Gamboa, G., Garde-Cerdán, T., Rubio-Bretón, P. & Pérez-Álvarez, E. P. (2020). Biostimulation to Tempranillo grapevines (Vitis vinifera L.) through a brown seaweed during two seasons: Effects on grape juice and wine nitrogen compounds. *Scientia Horticulturae, 264,* 109177. doi:10.1016/j.scienta.2020.109177.

Gutiérrez-Gamboa, G., Zamudio, D., Stefanello, S., Tassinari, A. & Brunetto, G. (2022). Application of foliar urea to grapevines: productivity and flavour components of grapes. *Australian Journal of Grape and Wine Research, 28*, 27–40.

Gutiérrez-Gamboa, G., Zamudio, L. Stefanello, L. & Brunetto, G. (2022). Application of foliar urea to grapevines: productivity and flavour components of grape. *Australian Journal of Grape and Wine Research, 28,* 27–40.

Gutiérrez-Gamboa, G., Alañon-Sanchez, N., Mateluna-Cuadra, R. & Verdugo-Vasquez, N. (2020). An overview about the impacts of agricultural practices on grape nitrogen composition: current research approaches. *Food Research International, 136,* 109477.

Gutiérrez-Gamboa, G., Romanazzi, G., Garde-Cerdan, T. & Pérez Alvarez, E. P. (2019). A review of the use of biostimulants in the vineyard for improved grape and wine quality: effects on prevention of grapevine diseases. *Journal of the Science of Food and Agriculture, 99,* 1001–1009.

Hanlin, R. L. & Downey, M. O. (2009). Condesend tanin accumulation and composition in skin of Shiraz and Cabernet Sauvignon grapes during berry development. *American Journal of Enology and Viticulture, 60*, 13-23.

Hannam, K. D., Neilsen, G. H., Neilsen, D., Midwood, A. J., Millard, P., Zhang, Z., Thornton, B. & Steinke, D. (2016). Amino acid composition of grape (Vitis vinifera L.) juice in response to applications of urea to the soil or foliage. *American Journal of Enology and Viticulture, 67,* 47–55.

Hannan, J. M. (2011). Potassium-Magnesium Antagonism in High Magnesium Vineyard Soils. Master's thesis, Iowa State University, Ames IA.

Harbertson, J. F., Hodgins, R. E., Thurston, L. N., Schaffer, L. J., Reid, M. S. & Landon, J. L. (2008). Variability of tannin concentration in red wine. *American Journal of Enology and Viticulture, 59,* 210-214.

Harborne, J. B. & Williams, C. A. (2001). Anthocyanins and other flavonoids. *Natural Product Report, 18,* 310-333.

Harborne, J. B. & Simmonds, N. W. (1964). *Biochemistry of Phenolic Compounds,* Academic Press, London, 101.

Havlin, J. L., Austin, R., Hardy, D., Howard, A. & Heitman, J. L. (2022). Nutrient Management Effects on Wine Grape Tissue Nutrient Content. *Plants, 11,* 158. https://doi.org/10.3390/plants11020158.

Hayasaka, Y., Waters, E. J., Cheynier, V., Herderich, M. J., & Vidal, S. (2003). Characterization of proanthocyanidins in grape seeds using electrospray mass spectrometry. *Rapid communications in mass spectrometry, 17*(1), 9-16.

He, F., Mu, L., Yan, G. L., Liang, N. N., Pan, Q. H., Wang, J., Reeves, M. J. & Duan C. Q. (2010). Biosynthesis of anthocyanins and their regulation in colored grapes. *Molecules, 15*, 9057–9091.

Hidalgo-Togores, J. (2002). *Tratado de Enologia. Madrid (Spain):* Ediciones MundiPrensa.

Ho, L., Ferruzzi, M. G., Janle, E. M., Wang, J., Gong, B., Chen, T. Y., Lobo, J., Cooper, B., Wu, Q. L., Talcott, S. T., Percival, S. S., Simon, J. E. & Pasinetti, G. M. (2013). Identification of brain-targeted bioactive dietary quercetin-3-O-glucuronide as a novel intervention for Alzheimer's disease. *The FASEB Journal, 27*(2), 769–781.

Hocking, B., Tyerman, S. D., Burton, R. A. & Gilliham, M. (2016). Fruit calcium: transport and physiology. *Front. Plant Sci. 7*, 569. doi: 10.3389/fpls.2016.00569.

Holzapfel, B. P., Watt, J., Smith, J. P., Suklje, K. & Rogiers. S. Y. (2015). Effects of timing of N application and water constraints on N accumulation and juice amino N concentration in "Chardonnay" grapevines. *Vitis, 54*(4), 203–11. doi: 10.5073/VITIS.2015.54.203-211.

Holzapfel, B. P. & Treeby, M. T. (2007). Effects of timing and rate of N supply on leaf nitrogen status, grape yield and juice composition from Shiraz grapevines grafted to one of three different rootstocks. *Australian Journal of Grape and Wine Research, 13*(1), 14–22.

Hornedo-Ortega, R., González-Centeno, M. R., Chira, K., Jourdes, M. & Teissedre, P. L. (2020). Phenolic Compounds of Grapes and Wines: Key Compounds and Implications in Sensory Perception. In F. Cosme, F. M. Nunes, L. Filipe-Ribeiro (Eds.), *Chemistry and Biochemistry of Winemaking, Wine Stabilization and Aging*, London: IntechOpen, 1–27.

Hummes, A. P., Bortoluzzi, E. C., Tonini, V., da Silva, L. P. & Petry, C. (2019). Transfer of copper and zinc from soil to grapevine-derived products in young and centenarian vineyards. *Water, Air, & Soil Pollution, 230*(7), 1–11. doi: 10.1007/s11270-019-4198-6.

Hussein, S. & Abdrabba, S. (2015). Physico-chemical Characteristics, Fatty Acid, Composition of Grape Seed Oil and Phenolic Compounds of Whole Seeds, Seeds and Leaves of Red Grape in Libya. *International Journal of Applied Science and Mathematics, 2*(5), 2394-2894.

Jackson, R. (2008). *Wine science-principles and aplications*, third edition, Elsevier.

Jackson, S. J. (2008). Wine science principles and applications, Third edition, Academic press.

James, A., Mahinda, A., Mwamahonje, A., Rweyemamu, E., Mrema, E., Aloys, K., Swai, E., Mpore, F. & Massawe, C. (2022). A review on the influence of fertilizers application on grape yield and quality in the tropics. *Journal of Plant Nutrition,* 46(12), 2936-2957. doi: 10.1080/01904167.2022.2160761.

Jegadeeswari, T., Chitdeshwari, A. K. & Shukla, D. (2020). Effect of multi-nutrients application on the yield and quality of grapes variety muscat. *Journal of Experimental Biology and Agricultural Sciences, 8*(4), 426–433. doi: 10. 18006/2020.8(4).426.433.

Johnson, H. (1989). *The story of wine,* Mitchell Beaszley, London.

Jones-Moore, H. R., Jelley, R. E., Marangon, M. & Fedrizzi, B. (2022). The interactions of wine polysaccharides with aroma compounds, tannins, and proteins, and their importance to winemaking, *Food Hydrocolloids, 123*, 107150. https://doi.org/10. 1016/j.foodhyd.2021.107150.

Kandylis, P., Dimitrellou, D. & Moschakis, T. (2021). Recent applications of grapes and their derivatives in dairy products. *Trends in Food Science & Technology, 114*, 696–711. doi: 10.1016/j.tifs.2021.05.029.

Karagiannidis, N., Nikolaou, N., Ipsilantis, I. & Zioziou, E. (2007). Effects of different N fertilizers on the activity of Glomus mosseae and on grapevine nutrition and berry composition. *Mycorrhiza, 18*(1), 43–50. doi: 10.1007/S00572-007-0153-2.

Karimi, P., Saberi, A. & Khadivi, A., (2021). Effects of foliar spray of agricultural grade mineral oil in springtime, in combination with potassium and calcium sulfates on the phenological and biophysical indices of clusters, and foliar nutritional levels in grapevine (Vitis vinifera L.) cv. Sultana (Id. Thompson seedless, Sultanina). *Biological Research, 54*, 28. https://doi.org/10.1186/s40659-021-00353-3.

Karimi, R. (2020). The effect of early season nutrition of calcium and zinc on yield, sugar content and enzymatic and non-enzymatic antioxidant capacity of grape. *Iran J Plant Bio., 12,* 1–22. https://doi.org/10.22108/ijpb.2019. 117399.1157.

Katalinić, V., Smole Mozina, S., Generalić, I., Skroza, D., Ljubenkov, I. & Klančnik, A. (2013). Phenolic profile, antioxidant capacity, and antimicrobial activity of leaf extracts from six Vitis vinifera L. varieties. *International Journal of Food Properties, 16*, 45-60.

Keller, M. (2010). *The science of grapevines-anatomy and physiology.* Elsevier.

Keller, M. (2015). *The science of grapevines: anatomy and physiology.* Burlington: Academic Press, 400.

Keller, M. (2020). *The science of grapevines,* 3d ed. Academic Press: Cambridge, MA, USA.

Kennedy, J. A., Saucier, C. & Glories, Y. (2006). Grape and wine phenolics: History and perspective. *American Journal of Enology and Viticulture, 57*(3), 239-248.

Khoshgoftarmanesh, A. H., Schulin, R., Chaney, R. L., Daneshbakhsh, B. & Afyuni, M. (2010). Micronutrient-efficient genotypes for crop yield and nutritional quality in sustainable agriculture: A review. *Agron. Sustain. Dev., 30*, 83–107.

Korać, N., Cindrić, P., Medić, M. & Ivanišević, D. (2016). *Viticulture and viticulture (part of viticulture)*. University of Novi Sad, Faculty of Agriculture, Novi Sad.

Korchagin, J., Moterle, D. F., Escosteguy, P. A. V. & Bortoluzzi, E. C. (2020). Distribution of copper and zinc frac tions in a Regosol profile under centenary vineyard. *Environmental Earth Sciences, 79*(19), 1–13. doi: 10.1007/s12665-020-09209-7.

Koslitz, S., Renaud, L., Kohler, M. & Wüst, M. (2008). Stereo selective formation of the varietal aroma compound rose oxide during alcoholic fermentation. *Journal of Agricultural and Food Chemistry, 56*, 1371–1375.

Košmerl, T., Jakončič, M., Kralj Cigić, I., Strlič, M. & Prosen, H. (2008). Aroma compound in „sur lies" produced and aged Chardonay wines. *31th Congesso mondiale della vigna e del vino*, 6a Assemblea Generale dell OIV, Verona.

Krol, A., Amarowicz, R. & Weidner, S. (2014). Changes in the composition of phenolic compounds and antioxidant properties of grapevine roots and leaves (Vitis vinifera L.) under continuous of long-term drought stress. *Acta Physiologiae Plantarum, 36*, 1491-1499.

Krus, T. E. C., Yu, Z., Pretson, C. M., Dahlgren, R. A. & Zasoski, R. J. (2003). Linking chemical reactivity and protein precipitation to structural characteristics of foliar tannins. *Journal of Chemical Ecology, 29*, 703-730.

Kumar, S., Kumar, S., & Mohapatra, T. (2021). Interaction between macro- and micro-nutrients in plants. *Frontiers in Plant Science, 12,* 665583. doi: 10.3389/FPLS.2021.665583/BIBTEX.

Lachman, J., Šulc, M., Faitová, K. & Pivec, V. (2009). Major factors influencing antioxidant contents and antioxidant activity in grapes and wines. *International Journal of Wine Research, 1,* 101–121.

Lai, H. Y., Juang, K. W. & Chen, B. C. (2010). Copper concentrations in grapevines and vineyard soils in central Taiwan. *Soil Sci Plant Nutr, 56,* 601–606. doi:10.1111/j.1747-0765.2010.00494.

Lasa, B., Menendez, S., Sagastizabal, K., Cervantes, M. E. C., Irigoyen, I., Muro, J., Aparicio-Tejo, P. M. & Ariz, I. (2012). Foliar application of urea to "Sauvignon Blanc" and "Merlot" vines: Doses and time of application. *Plant Growth Regulation, 67*(1), 73–81.

Leeuwen, C., Barbe, J. C., Geffroy, O., Gowdy, M., Lytra, G., Pons, A., Thibon, C. & Marchand, S. (2023). How terroir shapes aromatic typicity in grapes and wines (Part I). Sourced from the research article: "Recent advancements in understanding the terroir effect on aromas in grapes and wines." *OENO One, 2020. Technical reviews.* https://doi.org/10.20870/IVES-TR.2023.735.

Lena do Nascimento Silva, F., Schmidt, E. M., Messias, C. L., Eberlin, M. N. & Frankland Sawaya, H. A. C. (2015). Quantitation of organic acids in wine and grapes by direct infusion electrospray ionization mass spectrometry. *Analytical Methods, 7*(1), 53–62. doi:10.1039/c4ay00114a.

Lenk, S., Buschmann, C. & Pfündel, E. E. (2007). In vivo assessing flavonols in white grape berries (Vitis vinifera L. Cv Pinot Blanc) of different degrees of ripeness using chlorophyll fluoroscence imaging. *Functional Plant Biology, 34,* 1092-1104.

Likar, M., Vogel-Mikus, K., Potisek, M., Hancevic, K., Radic, T., Necemer, M. & Regvar, M. (2015). Importance of soil and vineyard management in the determination of grapevine mineral composition. *The Science of the Total Environment, 505*, 724–31. doi: 10.1016/J.SCITOTENV.2014.10.057.

Liu, Y., Pan, Q., Yan, G., He, G. & Duan, C. (2010). Changes of Flavan-3-ols with Different Degrees of Polimerization in Seeds of „Shiraz," „Cabernet Sauvignon" and „Marselan" Grapes after Veraison. *Molecules, 15,* 7763-7774.

Longbottom, M. (Ed). (2009). *Managing grapevine nutrition in a changing environment, Research to Practice Manual.* Australian Wine Research Institute, Adelaide, South Australia.

Lorensini, F., Ceretta, C. A., Girotto, E., Cerini, J. B., Lourenzi, C. R., De Conti, L., Trindade, M. M., de Melo, G. W. & Brunetto, G. (2012). Lixiviaçao e volatilizaçao de nitrogênio em um Argissolo cultivado com videira submetida à adubaçao nitrogenada. *Ciência Rural, 42,* 1173–1179.

Luiz, M. (2011). Proanthocyanidin profile and antioxidant capacity of Brazilian Vitis vinifera red wines, *Food Chemistry, 126*, 213–220.

Macheix, J. J., Fleuriet, A. & Billot, J. (1990). *Fruit Phenolics.* CRC Press, Florida, Boca Raton.

Malićanin, M. (2014). *Isolation and physicochemical characterization of oil from the seeds of red grape varieties.* University of Belgrade, Serbia.

Marković, N. (2012). *Technology of growing vines.* Monograph, Endowment of St. Hilandar Monastery, Belgrade.

Martinello, M., Hecker, G. & Pramparo, M. C. (2007). Grape seed oil deacidification by molecular distillation: analysis of operative variables influence using the response surface methodology. *Journal of Food Engineering, 81*, 60–64.

Martínez-Huélamo, M., Rodríguez-Morató, J., Boronat, A. & De la Torre, R. (2017). Modulation of Nrf2 by Olive Oil and Wine Polyphenols and Neuroprotection. *Antioxidants, 6(4),* 73. https://doi.org/10.3390/antiox6040073.

Martins, V., Teixeira, A., Bassil, E., Blumwald, E. & Gerós, H. (2014). Metabolic changes of Vitis vinifera berries and leaves exposed to Bordeaux mixture. *Plant Physiology and Biochemistry, 82,* 270-278. https://doi.org/10.1016/j.plaphy.2014.06.016.

Masi, E. & Boselli, M. (2011). Foliar application of molybdenum: Effects on yield quality of the grapevine sangio vese (Vitis vinifera L.). *Advances in Horticultural Science, 25*(1), 37–43. doi: 10.1400/172045.

Mattivi, F., Guzzon, R., Vrhovsek, U., Stefanini, M. & Velasco, R. (2006). Metabolite profiling of grape: flavonols and anthocyanins. *Journal of agricultural and food chemistry, 54*(20), 7692-7702.

Mattivi, F., Vrhovsek, U., Masuero, D. & Trainotti, D. (2009). Differences in amount and structure of extractable skin and seed tannins amongst red grape varieties. *Australian Journal of Grape and Wine Research, 15,* 27-35.

Matus, J. T., Loyola, R., Vega, A., Peña-Neira, A., Bordeu, E. & Arce-Johnson, P. et al., (2009). Post-veraison sunlight exposure induces MYB-mediated transcriptional regulation of anthocyanin and flavonol synthesis in berry skins of Vitis vinifera. *J. Exp. Bot. 60,* 853–867. doi: 10.1093/jxb/ern336.

Mendes-Pinto, M. M. (2009). Carotenoid breakdown products the–norisoprenoids–in wine aroma. *Arch. Biochem. Biophys., 483*, 236-245.

Mengel, K. (2007). Potassium. In: Barker A. V., Pilbeam D. J., editors. *Handbook of plant nutrition.* Boca Raton: CRC Press, 91–120.

Michopoulos, P. & Solomou, A. (2019). Effects of conventional and organic (manure) fertilization on soil, plant tissue nutrients and berry yields in vineyards. The use of the original native soil as a control. *Journal of Plant Nutrition, 42*(18), 2287–2298. doi: 10.1080/01904167.2019.1656246.

Milosavljević, M. (1984). Carbon uptake and metabolism in grapevines. In: *Physiology of the vine, M. Sarić*, Serbian Academy of Sciences, Belgrade, 25–50.

Miotto, A., Ceretta, A., Brunetto, B., Nicoloso, F., Girotto, E., Farias, J., Tiecher, T., Conti, L., & Trentin, G. (2014). Copper uptake, accumulation and physiological changes in adult grapevines in response to excess copper in soil. *Plant Soil, 374,* 593–610. doi: 10.1007/s11104-013-1886-7.

Mitic, M., Jankovic, S., Mitic, S., Kocic, G., Maskovic, P. & Djukic, D. (2021). Optimization and Kinetic Modelling of Total Phenols and Flavonoids Extraction from Tilia cordata M. Flowers. *S. Afr. J. Chem., 75*, 64–72.

Monagas, M., Gómez-Cordovés, C., Bartolomé, B., Laureano, O. & Ricardo da Silva, J. M. (2003). Monomeric, oligomeric, and polymeric flavan-3-ol composition of wines and grapes from Vitis vinifera L. cv. Graciano, Tempranillo, and Cabernet Sauvignon. *J. Agric. Food Chem., 51*, 6475-6481.

Monagas, M., Garrido, I., Bartolomé, B. & Gómez-Cordovés, C. (2006). Chemical characterization of commercial dietary ingredients from Vitis vinifera L. *Analytica chimica acta, 563*(1-2), 401-410.

Moyer, M., Singer, S., Davenport, J. & Hoheisel, G. (2018). *Vineyard nutrient management in Washington state*. Washington State University Extension and the U.S. Department of Agriculture.

Müller, C. & Riederer, M. (2005). Plant surface properties in chemical ecology. *Journal of Chemical Ecology, 31,* 2621–2651.

Mustafa, S. & Mustafa, S., (2023). Effect of Tipping and Foliar Application of Proline and Botminn Plus on Yield and Quality of Grapevine (Vitis vinifera L.) cv. Khoshnaw. Fifth International Conference for Agricultural and Environment Sciences. IOP Publishing. *IOP Conf. Series: Earth and Environmental Science* 1158, 042071 doi:10.1088/1755-1315/1158/4/042071.

Navarro, J. P., Ros, A., Masuero, D., Izquierdo-Cañas, P. M., Gutiérrez, I. H., Gómez-Alonso, S., Mattivi, F. & Vrhovsek, U. (2019). LC-MS/MS analysis of free fatty acid composition and other lipids in skins and seeds of Vitis vinifera grape cultivars. *Food Research International, 125,* 108556. https://doi.org/10.1016/j.foodres.2019.108556.

Nowshehri, J. A., Bhat, Z. A. & Shah, M. Y. (2015). Blessings in disguise: Bio-functional benefits of grape seed extracts. *Food Reviews International, 77,* 333-348.

O'Keefe, J. H., Bhatti, S. K., Bajwa, A., DiNicolantonio, J. J. & Lavie, C. J. (2014). Alcohol and Cardiovascular Health: The Dose Makes the Poison or the Remedy. *Mayo Clinic Proceedings, 89*(3), 382–393.

Ono, K., Yoshiike, Y., Takashima, A., Hasegawa, K., Naiki, H. & Yamada, M. (2003). Potent antiamyloidogenic and fibril-destabilizing effects of polyphenols in vitro: implications for the prevention and therapeutics of Alzheimer's disease. *Journal of Neurochemistry, 87*(1), 172–181.

Ortega-Regules, A., Romero-Cascales, I., Ros-Garcia, J. M., Lopez-Roca, J. M. & Gomez-Plaza, E. (2006). A first approach towards the relationship between grape skin cell-wall composition and anthocyaninextractability. *Analytica Chimica Acta, 563*, 26-32.

Passos, C. P., Silva, R. M., Da Silva, F. A., Coimbra, M. A. & Silva, C. M. (2009). Enhancement of the supercritical fluid extraction of grape seed oil by using enzymatically pre-treated seed. *The Journal of Supercritical Fluids, 48*(3), 225-229.

Pastrana-Bonilla, E., Akoh, C. C., Sellappan, S. & Krewer, G. (2003). Phenolic content and antioxidant capacity of muscadine grapes. *J. Agric. Food Chem.,* 51, 5497–4503.

Paul Schreiner, R., Osborne, J. & Skinkis, P. A. (2018). Nitrogen requirements of pinot noir based on growth parameters, must composition, and fermentation behavior. *American Journal of Enology and Viticulture, 69*(1), 45–58. doi: 10.5344/ajev.2017.17043.

Peacock, W. L. & Christensen, L. P. (2005). Drip irrigation can effectively apply boron to San Joaquin Valley vine yards. *California Agriculture, 59*(3), 188–191. https://escholarship.org/uc/item/8fb93784. doi: 10.3733/ca. v059n03p188.

Pereira, D. M., Valentao, P., Pereira, J. A. & Andrade, P. B. (2009). Phenolics: From Chemistry to Biology. *Molecules, 14*, 2202-2211.

Perez-Alvarez, E. P., Garde-Cerdan, T., Garcıa-Escudero, E. & Martınez-Vidaurre, J. M. (2017). Effect of two doses of urea foliar application on leaves and grape nitrogen composition during two vintages. *Journal of the Science of Food and Agriculture, 97*(8), 2524–2532. doi: 10.1002/JSFA.8069.

Pezet, R., Perret, C., Jean-Denis, J. B., Tabacchi, R., Gindro, K. & Viret, O. (2003). δ-Viniferin, a resveratrol dehydrodimer: one of the major stilbenes synthesized by stressed grapevine leaves. *Journal of agricultural and Food Chemistry, 51*(18), 5488-5492.

Pezzuto, J. M. (2008). Grapes and human health: A perspective. *Journal of Agricultural and Food Chemistry, 56*(16), 6777–6784. doi: 10.1021/JF800898P.

Pinelo, M., Arnous, A. & Meyer, A. S. (2006). Upgrading of grape skins: significance of plant cell-wall structural components and extraction techniques for phenol release. *Trends in Food Science & Technology, 17,* 579–590.

Pomar, F., Novo, M. & Masa, A. (2005). Varietal differences among the anthocyanin profiles of 50 red table grape cultivars studied by high performance liquid chromatography. *Journal of Chromatography A, 1094*, 34-41.

Porro, D., Policarpo, M., Stefanini, M., Dorigatti, C., Ziller, L. & Camin, F. (2010). Nitrogen foliar uptake and partitioning in “Cabernet Saugvinon” Grapevines. *Acta Hortic., 868,* 185-190. https://doi.org/10.17660/ActaHortic.2010.868.21.

Proffitt, T. & Campbell-Clause, J. (2021). *Managing Grapevine Nutrition and Vineyard Soil Health Perth Region NRM.* Available online: www.winewa.asn.au.

Prosen, H., Janeš, L., Strlič, M., Rusjan, D. & Kočar, D. (2007). Analysis of free and bound aroma compounds in grape berries using headspace solid-phase microextraction with GC-MS and a preliminary study of solid-phase extraction with LC-MS. *Acta. Chim. Slov., 54,* 25-32.

Pržić, Z. (2015). The influence of defoliation on the content of the most important compounds of the aromatic and flavonoid complex in grapes and wine of some grape varieties. University of Belgrade. *Doctoral dissertation.*

Puskaš, V. (2010). Influence of technological factors in the production of red wines on the content and stability of catechins and their oligomers, *Doctoral dissertation,* University of Novi Sad.

Rahaman, D. M. M., Baby, T., Oczkowski, A., Paul, M., Zheng, L., Schmidtke, L. M., Holzapfel, B. P., Walker, R. R. & Rogiers, S. Y. (2019). Grapevine nutritional disorder detection using image processing. *Lecture Notes in Computer Science, 11854*, 184–96. doi: 10.1007/978-3-030-34879-3_15/COVER/.

Ramos, S. (2008). Cancer chemoprevention and chemotherapy: dietary polyphenols and signalling pathways. *Molecular Nutrition and Food Research, 52*, 507–526.

Ramsey, R. J. L., Stephenson, G. R. & Hall, J. C. (2005). A review of the effects of humidity, humectants, and surfactant composition on the absorption and efficacy of highly water-soluble herbicides. *Pesticide Biochemistry and Physiology, 82*, 162–175.

Rawson, G. A. (2002). *The influence of geology and soil characteristics on the fruit composition of winegrape (Vitis Vinifera cv. Shiraz), Hunter valley, New South Wales: Implications for regionality in the Australian wine industry*. University of Newcastle, Newcastle, NSW, 273.

Reis Giada, M. de L. (2013). Food Phenolic Compounds: Main Classes, Sources and Their Antioxidant Power. *InTech*. doi: 10.5772/51687.

Reynolds, B. H. J. (eds.). (2022*). Wheat Improvement. School of Agriculture Food and Wine*, University of Adelaide, Adelaide, SA, Australia.

Ribéreau-Gayon, P. (1972). Plant Phenolics. In: Hegwood, V. H. (Ed) *Plant Phenolics, University review in botany.* Oliver and Boyd, Edinburgh, 254.

Ribereau-Gayon, P., Glories, Y., Maujean, A. & Dubourdieu, D. (2006). *Handbook of enology 2,* The chemistry of Wine Stabilization and Treatments. John Wiley & Sons Ltd. England.

Rinaldi, A., Gonzalez, A., Moio, L. & Gambuti, A. (2021). Commercial Mannoproteins Improve the Mouthfeel and Colour of Wines Obtained by Excessive Tannin Extraction. *Molecules, 26*(14), 4133. https://doi.org/10.3390/molecules26144133.

Robbins, R. J. (2003). Phenolic Acids in Foods: An Overview of Analytical Methodology. *Journal of Agricultural and Food Chemistry, 51*, 2866–2887.

Rodrigo, R., Miranda, A. & Vergara, L. (2011). Modulation of endogenous antioxidant system by wine polyphenols in human disease. *Clinica Chimica Acta, 412*(5-6), 410-424.

Rogiers, S. Y., Coetzee, Z. A., Walker, R. R., Deloire, A. & Tyerman, S. D. (2017). Potassium in the grape (Vitis vin ifera L.) berry: Transport and function. *Frontiers in Plant Science, 8*, 1629. doi: 10.3389/FPLS.2017.01629/BIBTEX.

Rogiers, S., Zheng, L., Oczkowski, A., Rahaman A., Paul, M., Tintu Baby, T., Mentjox, K., Schmidtke., Walker, R. & Holzapfel, B. (2021). Vine Nutrition: A Diagnostic App for Nutrient Disorder Assessment. *Final report to wine Australia.*

Rogiers, S. Y., Hardie, W. J. & Smith, J. P. (2011). Stomatal density of grapevine leaves (Vitis vinifera L.) responds to soil temperature and atmospheric carbon dioxide. *Australian Journal of Grape and Wine Research, 17*, 147–152.

Romic, M., Zovko, M., Romic, D. & Bakic, H. (2012). Improvement of vineyard management of Vitis vinifera L. cv. Grk in the Lumbarda Vineyard Region (Croatia). *Communications in Soil Science and Plant Analysis, 43*(1-2), 209–218. doi: 10.1080/00103624.2011.638557.

Rustioni, L., Grossi, D., Brancadoro, L. & Failla, O. (2017). Characterization of iron deficiency symptoms in grapevine (Vitis spp.) leaves by reflectance spectroscopy. *Plant Physiology and Biochemistry, 118,* 342e347.

Sanchez-Palomo, E., Diaz-Maroto, H. M. C., Gonzales-Vinas, M. A., Soriano-Perez, A. & Perez-Coelo, M. S. (2007). Aroma profile of wines from Albillo and Muscat grape varieties at different stages of ripening. *Food Control., 18*, 398-403.

Savocchia, S., Mandel, R., Crisp, P. & Scott, E. S. (2011). Evaluation of 'alternative' materials to sulfur and synthetic fungicides for control of grapevine powdery mildew in a warm climate region of Australia. *Australasian Plant Pathology, 40*(1), 20–27. doi: 10.1007/s13313-010-0009-7.

Schreiner, R. P. & Osborne, J. (2018). Defining phosphorus requirements for pinot noir grapevines. *American Journal of Enology and Viticulture, 69*(4), 351–359. doi: 10.5344/ajev.2018.18016.

Schreiner, R. P., Lee, J. & Skinkis, P. A. (2013). N, P, and K supply to pinot noir grapevines: Impact on vine nutri ent status, growth, physiology, and yield. *American Journal of Enology and Viticulture, 64*(1), 26–38. doi: 10. 5344/ajev.2012.12064.

Schreiner, R. P., Scagel, C. F. & Lee, J. (2014). N, P, and K Supply to Pinot noir Grapevines: Impact on berry phenolics and free amino acids. *American Journal of Enology and Viticulture, 65*(1), 43–49. doi: 10.5344/ajev.2013. 13037.

Schwab, W., Davidovich-Rikanati, R. & Lewinsohn, E. (2008). Biosynthesis of plant derived flavour compounds. *The Plant Journal, 54*, 712-732.

Scutarașu, C. E., Luchian, E. C., Colibaba, C. L. & Cotea, V. (2023). Enzymes and Biochemical Catalysis in Enology: Classification, Properties, and Use in Wine Production. *IntechOpen*. doi: 10.5772/intechopen.105474.

Serrano, J., Silva, J. M. da., Shahidian, S., Silva, L. L., Sousa, A. & Baptista, F. (2017). Differential vineyard fertilizer management based on nutrients spatio-temporal variability. *Journal of Soil Science and Plant Nutrition, 17*. doi: 10.4067/S0718-95162017005000004.

Silva, P. & Vauzour, D. (2018). Wine Polyphenols and Neurodegenerative Diseases: An Update on the Molecular Mechanisms Underpinning Their Protective Effects. *Beverages*, *4*(4), 96. https://doi.org/10.3390/beverages4040096.

Souqet, J. M., Labarbe, B., Le Guerneva, C., Cheynier, V. & Moutounet, M. (2000). Phenolic composition of grape stems. *Journal of agricultural and Food Chemistry, 48*, 1076-1080.

Spayd, S. E., Tarara, J. M., Mee, D. L. & Ferguson, J. C. (2002). Separation of sunlight and temperature effects on the composition of Vitis vinifera cv. Merlot berries. *American Journal of Enology and Viticulture, 53*(3), 171-182.

Stefanello, L. O., Schwalbert, R., Schwalbert, R. A., De Conti, M. S. Kulmann, S., Garlet, L. P., Silveira, M. L. R., Sautter, C. K., Melo, G. B. W. & Rozane, D. E. et al., (2020). Nitrogen supply method affects growth, yield and must composition of young grape vines (Vitis vinifera L. cv Alicante Bouschet) in southern Brazil. *Scientia Horticulturae, 261*, 108910. doi: 10.1016/j.scienta.2019.108910.

Stefanello, L. O., Schwalbert, R., Schwalbert, R. A., Drescher, G. L., Conti, L. De., Pott, L. P., Tassinari, A., Kulmann, M. S. & Silva, S. I. C. B. et al., (2021). Ideal nitrogen concentration in leaves for the production of high quality grapes cv 'Alicante Bouschet' (Vitis vinifera L.) subjected to modes of application and nitrogen doses. *European Journal of Agronomy, 123*, 126200. doi: 10.1016/j.eja.2020.126200.

Stefanello, L. O., Schwalbert, R., De-Conti, L., Tassinari, A., Paula Garlet, L., Lourenzi, C. R., Comin, J., Loss, A., Schmitt, D., Borghezan, M., Ambrosini, V. & Brunetto, G. (2019). Yield and must composition of "Cabernet Sauvignon" grapevines subjected to nitrogen application in soil with high organic matter content. *Idesia, 37*, 27–36.

Stefanello, L. O., Schwalbert, R., Schwalbert, R. A., De Conti, L., Kulmann, M. S. D. S., Garlet, L. P., Silveira, M. L. R., Sautter, C. K., de Melo, G. W. B., Rozane, D. E. &

Brunetto, G. (2020a). Nitrogen supply method affects growth, yield and must composition of young grape vines (Vitis vinifera L. cv Alicante Bouschet) in south ern Brazil. *Scientia Horticulturae, 261*, 108910.

Stines, A. P., Grubb, J., Gockowiak, H., Henschke, P. A., Hoj P. B. & van Heeswijck R. (2000). Proline and arginine accumulation in developing berries of Vitis vinifera L. in Australian vineyards: influence of vine cultivar, berry maturity and tissue type. *Australian Jour. of Grape & Wine Research, 6,* 150-158.

Stojanova, T. M. (2018). *Plant nutrition*. Academic Press, Skopje.

Stojanova, T. M. (2019). *Manual for taking soil and plant samples for agrochemical analysis.* Academic Press, Skopje.

Stojanova, T. M. (2020). *Fruits nutrition*. Academic Press, Skopje.

Stojanova, M., Djukic, D., Stojanova, M. T., Şatana, A. & Lalevic, B. (2023b). Antioxidant and Antimicrobial Potential of Wild Chestnut *Ohrid Diamond* Extract – An Opportunity for Creation of New Natural Products. *Erwerbs-Obstbau, 65*(1), 83-91. https://doi.org/10.1007/s10341-022-00668-9.

Stojanova, M., Dragutin, A. Djukic., Stojanova T. M. & Şatana, A. (2024). Modern Extraction Methods of Biologically Active Components in Food Biotechnology. *Monography.* Cambridge Scholars Publishing, UK.

Stojanova, M., Đukić, D. & Stojanova, M. T. (2023). Possibilities of using Coriolus versicolor lyophilized extracts in the industrial production of soups. *International Paris Congress on Applied Sciences. Proceedings Book*, IKSAD Institute, Paris, France, 86–93.

Stojanova, M., Pantić, M., Karadelev, M., Čuleva, B. & Nikšić, M. (2021). Antioxidant potential of extracts of three mushroom species collected from the Republic of North Macedonia. *Journal of Food Processing and Preservation, 45*(2), e15155. https://doi.org/10.1111/jfpp.15155.

Stojanova, M., Trpeski, V. & Boshkov, K. (2005). The influence of mineral fertilization on the content of phosphorus and potassium in the leaves of Chardonnay and Italian Riesling variety in the Skopje vineyard. *Annual proceedings of Faculty of Agricultural Sciences and Food, Skopje, 50*, 37–44.

Stojanova, M., Trpeski, V., Boskov, K. & Knevezic, B. (2007). The influence of mineral fertilizing on the chemical composition of must from Chardonnay and Italian Riesling varieties. *Jubilee Annual Proceedings of Faculty of Agricultural Sciences and Food, Skopje, 52*, 117–123.

Stojanova, T. M. (2023). *Nutrition of the grapevine*. Academic thought, Skopje.

Stojanova, T. M., Djukic, D. & Stojanova, M. (2023c). The effect of foliar fertilizing on the plum yield and the chemical composition of plum leaves. *7th International Conference on Innovative Studies of Contemporary Sciences, Proceedings Book,* Iksad Institute Publishing House, Tokyo, Japan.

Stojanova, T. M., Ivanovski, I., Stojkova, I. & Stojanova, M. (2016). The effect of foliar fertilizing on the chemical composition of kernels of primorski almond cultivar grown in Valandovo. *International Journal of Current Advanced Research, 5*(5), 891-894.

Sun, B., Ribes, A. M., Leandro, M. C., Belchior, A. P. & Spranger, M. I. (2006). Stilbenes: quantitative extraction from grape skins, contribution of grape solids to wine and variation during wine maturation. *Analytica Chimica Acta, 563*(1-2), 382-390.

Swain, T. & Bate-Smith, E. C. (1962). Flavonoid compounds, in: *Comparative Biochemistry* Vol. III., M. Florkin and H. S. Mason (eds.), Academic Press, New York, NY, 755–809.

Sweetman, C., Wong, D. C., Ford, C. M. & Drew, D. P. (2012). Transcriptome analysis at four developmental stages of grape berry (Vitis vinifera cv. Shiraz) provides insights into regulated and coordinated gene expression. *BMC Genomics, 13*, 691. doi: 10.1186/1471-2164-13-691.

Tanaka, Y., Sasaki, N. & Ohmiya, A. (2008). Biosynthesis of plant pigments: anthocyanins, betalains and carotenoids. *The Plant Journal, 54*, 733-749.

Tangolar, S., Tangolar, S., Turan, M., Atalan, M. & Ada, M. (2023). The Effects of Different Substrates with Chemical and Organic Fertilizer Applications on Vitamins, Mineral, and Amino Acid Content of Grape Berries from Soilless Culture. *IntechOpen.* doi: 10.5772/intechopen.102345.

Taware, P. B., Dhumal, K. N., Oulkar, D. P., Patil, S. H. & Banerjee, K. (2010). Phenolic alterations in grape leaves, berries and wines due to foliar and cluster powdery mildew infections. *International Journal of Pharma and Bio Sciences, 1*, 1-14.

Teixeira, A., Eiras-Dias, J., Castellarin, S. D. & Geros, H. (2013). Berry phenolics of grapevine under challenging environments. *International Journal of Molecular Sciences, 14*, 18711-18739.

Terral, J. F., Tabard, E., Bouby, L., Ivorra, S., Pastor, T. & Figueiral, I. (2010). Evolution and history of grapevine (Vitis vinifera) under domestication: new morphometric perspectives to understand seed domestication syndrome and reveal origins of ancient European cultivars. *Annals of Botany, 105,* 443–455.

Thomidis, T., Zioziou, E., Koundouras, S., Karagiannidis, C., Navrozidis, I. & Nikolaou, N. (2016). Effects of nitrogen and irrigation on the quality of grapes and the susceptibility to Botrytis bunch rot. *Scientia Horticulturae, 212*, 60-68. doi:10.1016/j. scienta.2016.09.036.

Tian, R. R., Pan, Q. H., Yhen, J. Ch., Li, J. M., Wan, S. B., Yhang, Q. H. & Huang. W. D. (2009). Comparison of phenyl acids and flavan-3-ols during wine fermentation of grapes with different harvest times. *Molecules, 14*, 827-838.

Tian, T., Ruppel, M., Osborne, J., Tomasino, E. & Schreiner, R. P. (2022). Fertilize or supplement: The impact of nitrogen on vine productivity and wine sensory properties in Chardonnay. *American Journal of Enology and Viticulture, 73*(3), 156–169. doi: 10.5344/ajev.2022.21044.

Truelsen, T., Thudium, D. & Gronbaek, M. (2002). Amount and type of alcohol and risk of dementia: The Copenhagen City Heart Study. *Neurology, 59*(9), 1313–1319.

Turner, R. S., Thomas, R. G., Craft, S., van Dyck, C. H., Mintzer, J., Reynolds, B. A., Brewer, J. B., Rissman, R. A., Raman, R. & Aisen, P. S. (2015). Alzheimer's Disease Cooperative Study. A randomized, double-blind, placebo-controlled trial of resveratrol for Alzheimer disease. *Neurology, 85*(16), 1383–1391.

Vepsäläinen, S., Koivisto, H., Pekkarinen, E., Mäkinen, P., Dobson, G., McDougall, G. J., Stewart, D., Haapasalo, A., Karjalainen, R. O., Tanila, H. & Hiltunen, M. (2013). Anthocyanin-enriched bilberry and blackcurrant extracts modulate amyloid precursor protein processing and alleviate behavioral abnormalities in the APP/PS1 mouse model of Alzheimer's disease. *The Journal of Nutritional Biochemistry, 24*(1), 360–370.

Verdenal, T., Dienes. A., Spangenberg, J., Zufferey, V., Spring, J., Viret, O., Marin-Carbonne, J. & van Leeuwen, C. (2020). Understanding and managing nitrogen nutrition in grapevine: a review. *OENO One, 55*(1), 1–43. https://doi.org/10.20870/oeno-one.2021.55.1.3866.

Verdenal, T., Dienes-Nagy, Ágnes, Spangenberg, J. E., Zufferey, V., Spring, J. L., Viret, O. & van Leeuwen, C. (2021). Understanding and managing nitrogen nutrition in grapevine: a review. *OENO One*, *55*(1), 1–43. https://doi.org/10.20870/oeno-one.2021.55.1.3866.

Verdenal, T., Spangenberg, J. E., Zufferey, V., Lorenzini, F., Dienes-Nagy, A., Gindro, K., Spring J. L. & Viret, O. (2016). Leaf-to-fruit ratio affects the impact of foliar applied nitrogen on N accumulation in the grape must. *Journal International des Sciences de la Vigne et du Vin, 50*(1), 23-33. doi:10.20870/oeno-one.2016.50.1.55.

Verdenal, T., Spangenberg, J. E., Zufferey, V., Lorenzini, F., Spring, J. L. & Viret, O. (2015). Effect of fertilisation timing on the partitioning of foliar-applied nitrogen in Vitis vinifera cv. Chasselas: a 15N labelling approach. *Australian Journal of Grape and Wine Research, 21*, 110–117.

Vermerris, W. & Nicholson, R. (2006). Families of phenolic compounds and means of classification. In: Vermerris W. and Nicholson, R. (eds.) *Phenolic Compound Biochemistry,* Springer, USA, 1−34.

Vinson, J. A., Su, X., Zubik, L. & Bose, P. (2001). Phenol antioxidant quantity and quality in foods: Fruits. *Journal of Agricultural and Food Chemistry, 49*, 5315–5321.

Wang, H., Liu, Y. M., Qi, Z. M., Wang, S. Y., Liu, S. X., Li, X., Wang, H. J. & Xia, X. C. (2013). An Overview on Natural Polysaccharides with Antioxidant Properties. *Current Medicinal Chemistry, 20,* 2899–2913. http://doi:10.2174/092986731132023006.

Wang, J., Ho, L., Zhao, Z., Seror, I., Humala, N., Dickstein, D. L., Thiyagarajan, M., Percival, S. S., Talcott, S. T. & Pasinetti, G. M. (2006). Moderate consumption of Cabernet Sauvignon attenuates Aβ neuropathology in a mouse model of Alzheimer's disease. *The FASEB Journal, 20*(13), 2313–2320.

White, P. J. & Brown, P. H. (2010). Plant nutrition for sustainable development and global health. *Annals of Botany, 105*(7), 1073–1080. doi: 10.1093/AOB/MCQ085.

Wilson, S. G., Lambert, J. J. & Dahlgren, R. (2021). Compost application to degraded vineyard soils: Effect on soil chemistry, fertility, and vine performance. *American Journal of Enology and Viticulture, 72*(1), 85–93. doi: 10. 5344/ajev.2020.20012.

Witheetrirong, Y., Tripathi, N. K., Tipdecho, T. & Parkpian, P. (2011). Estimation of the effect of soil texture on nitrate-nitrogen content in groundwater using optical remote sensing. *International Journal of Environmental Research and Public Health, 8*, 3416–3436.

Wood, J. E., Senthilmohan, S. T. & Peskin, A. V. (2002). Antioxidant activity of procyanidin-containing plant extracts at different pHs. *Food Chemistry, 77*(2), 155-161. https://doi.org/10.1016/S0308-8146(01)00329-6.

Woraratphoka, J., Intarapichet, K. O. & Indrapichate, K. (2007). Phenolic compounds and antioxidative properties of selected wines from the northeast of Thailand. *Food Chemistry, 104*(4), 1485-1490.

Youdim, K. A., McDonald, J., Kalt, W. & Joseph, J. A. (2002). Potential role of dietary flavonoids in reducing microvascular endothelium vulnerability to oxidative and inflammatory insults. *The Journal of nutritional biochemistry, 13*(5), 282-288.

Yruela, I., Alfonso, M., Barón, M. & Picorel, R. (2000). Copper effect on the protein composition of photosystem II. *Physiol Plant, 110*, 551–557. doi:10.1111/j.1399-3054.2000.1100419.x.

Yurchenko, E. & Artamonov, A. (2020). Technological Effectiveness of Chelated Micronutrient Fertilizers in Leaf Treatments Inducing Grapes Resistance to Biotic and Abiotic Stresses. *BIO Web of Conferences* 21, 00033. https://doi.org/10.1051/bioconf/20202100033.

Zebec, V., Lisjak, M., Jovic, J., Kujundzic, T., Rastija, D. & Loncaric, Z. (2021). Vineyard fertilization management for iron deficiency and chlorosis prevention on carbonate soil. *Horticulturae, 7*(9), 285. doi: 10.3390/horticulturae7090285.

Zhou, D. D., Li, J., Xiong, R. G., Saimaiti, A., Huang, S. Y., Wu, S. X., Yang, Z. J., Shang, A., Zhao, C. N. & Gan, R. Y. et al., (2022). Bioactive Compounds, Health Benefits and Food Applications of Grape. *Foods*, *11*(18), 2755. https://doi.org/10.3390/foods 11182755.

Zhu, L., Zhang, Y., Lu, J. (2012). Phenolic Contents and Compositions in Skins of Red Wine Grape Cultivars among Various Genetic Backgrounds and Originations. *Int. J. Mol. Sci., 13,* 3492-3510.

Zlámalová, T., Elbl, J., Baron, M., Belíková, H., Lampír, L., Hlušek, J. & Lošák, T. (2016). Using foliar applications of magnesium and potassium to improve yields and somequalitative parameters of vine grapes (Vitis vinifera L.). *Plant Soil Environ., 61,* 451–457.

About the Authors

Dr. Marina T. Stojanova is a full professor at the Faculty of Agricultural Sciences and Food, University of Ss. Cyril and Methodius, Skopje, Republic of North Macedonia. She is the head of the laboratory for agro-chemistry and plant nutrition. As an author and co-author, she has published more than 150 original research papers in peer-reviewed national and international journals. Additionally, Prof. Stojanova has presented and lectured at numerous symposiums, congresses and conferences in different countries such as Serbia, Turkey, France, Japan, China, and more. Moreover, she is the author of five books and two handbooks intended for students and farmers. She has participated in a large number of professional and educational projects.

E-mail: mstojanova@fznh.ukim.edu.mk.

Academician Dragutin A. Djukic is a microbiologist, and full professor at the Faculty of Agronomy, University of Kragujevac, Republic of Serbia. He published more than 400 scientific papers, 52 scientific announcements and 4 expert papers; 20 monographs, 6 textbooks, 4 practicums and 3 manuals. Prof. Djukic is one of the initiators of the establishment of the scientific-professional meeting "Review of Scientific Papers of Agronomy Students" and "Biotechnology Counselling" in Serbia. He is a member of the editorial board of the "Journal of Central European Agriculture," editor of the Journal "Acta Agriculturae Srebica" and a member of the editorial board of Journals "Bulgarian Agriculturae Science" and "Natura Montenegrina."

E-mail: djukicba@kg.ac.rs.

Monika Stojanova earned her undergraduate and master's degrees from the Faculty of Agricultural Sciences and Food at the University of Ss. Cyril and Methodius in Skopje, North Macedonia. She received the title of Doctor of Sciences at the Faculty of Agriculture, University of Belgrade, Serbia (in the field of food technology – industrial microbiology). Dr. Stojanova is the founder and president of the Association for Scientific-research, Educational and Cultural Activities "Open Science," North Macedonia. She has published more than 80 original research papers in peer-reviewed international journals, as well as in conference proceedings. With oral presentations, Stojanova has participated in numerous conferences in North Macedonia, Serbia, Bulgaria, Bosnia and Herzegovina, Japan, France, Azerbaijan and Turkey. Dr. Stojanova has authored two innovations: Bio-soups and Macedonian traditional sausage. At the World Fair of Women Innovators in Seoul, North Korea, she won two gold medals, one silver medal and a certificate for the best innovation. In addition, she has won numerous national and international awards and recognitions.

E-mail: stojanova.monika@yahoo.com.

Asst. Prof. Dr. Aziz Şatana holds an MSc and PhD from Trakya and Namık Kemal Universities in Turkey. He has also completed postdoctoral studies at North Dakota State University in the USA. He has participated in conferences across the globe, including in the USA, the Netherlands, Spain, Serbia, Bulgaria, Romania, Hungary, and Turkey. He was awarded a scholarship from the Spanish Government and has lectured in Germany, Hungary, and Bulgaria as part of the Erasmus Programme. In 2020, he was inducted as a member of the Royal Academy of Engineering in the UK and received an award. Currently, he works at Erciyes University in the Republic of Turkey as an adviser to the Turkish Agriculture Ministry and serves as the Chair of the WOCOLS Conference.

E-mail: azizsatana@hotmail.com.

Index

Note: Page numbers in *italics* indicate a figure and page numbers in **bold** indicate a table on the corresponding page.

A

accumulation of ions, factors affecting, 21–26
  endogenous, 21–24
  exogenous, 24–26
acetic acid, in wine, 99–100
acids, in grapes, 98–100
  citric, 99
  malic, 100
  organic, 100
  tartaric, 98
amino acids (AAs), 30–32, 43, 54–55, 68, 93, 104–105
analysis, 8, 13, 26, 28
anions, 15, 24, 28, 38–39, 102–103, 120
anthocyanins, 109–110, 119–124, 129–134, 139–144
  anthoxanthin, 135
  coloured substances in grapes, 129, 130, 139–140
  equilibrium forms, 134
  as glucosides, 122
  malvidin, 143
  structures of grape-derived, 121
  synthesis of, 120, 123, 130, 131, 133
  tannins and, 131, 141
antioxidant, 93–95, 110, 115, 117, 140–141, 144
aromatic compounds, 44, 96, 135–138
  muscat, 136
  primary, 136
  secondary, 136
  tertiary, 136
  varietal, 137

B

boron (B), 74
  deficiency, 74, 76–78, 88
  excess of, 74, 78, 88
  fertilizers, 74
  in nutrition of grapevine, 74
  symptoms of deficiency and excess, 78
  toxicity, 77, 81
  uptake by grapevines, 75
Botrytis cinerea, 47, 97, 99

C

calcium (Ca), 47–50
  auxin activity, 48
  deficiency, 48–50, 59
  excess of, 50, 60
  regulating permeability of cell membrane, 48
  role in grapevine, 47–50
  symptoms of deficiency and excess, 49–50
carbohydrates, in grapes, 96–98
  beta-D-glucose, 97
  sucrose structure, 97
carotenoids, 32, 69–70, 94–95, 106, 129
cations, 15, 48, 64, 68, 71, 102, 120
Chardonnay, 106, 137
chemical composition, of grapes, 92–108
  acids, 98–100

biological anatomy and, 94
carbohydrates, 96–98
carotenoids, 106
enzymes, 101–102
lipids, 105–106
mineral matters, 102–104
nitrogen compounds, 104–105
pectin substances, 107
vitamins, 107–108
volatile components, 107
waxes, 108
chlorosis, 28–29, 37, **52–53**, 66–68, 70–71, **72**
iron, 66, 71
cobalt (Co), 82
coloured substances, 129–135
of black grapes and red wines, 130–134
of red grapes, 130, 131
on white grapes, 135
copper (Cu), 71–74
deficiency, 73, 87
excess of, 73–74, 87
in grapevine, 71
and iron, 28
physiological function, 71
symptoms of deficiency and excess, 72

E

enzymes
action on grape pectin chains, 103
activity during ripening, 101–102
invertases, 101
oxido-reduction, 101
pectolytic, 101
epicatechin, 119, 124, *125*, 127, 146

F

fatty acids, 106, *106*, 108
fermentation52, 97, 99–100, 102, 104–105, 107
alcoholic, 36, 97, 99–100, 102, 104–105, 107
malolactic, 99–100
fertilization
foliar, 10–14, 24
soil, 10, 23
in viticulture production, 7–14
flavonoids, 114, 116–117, 124, 131, 139, 140, 144
anthocyanins, 110
biosynthesis, 102, 114
composition, 116–117
hydroxylase, 131
in skin and seeds, 116, 118
flavonols, 95, 109, 117, 122, 135, 141
in grapevine leaves, 114
quercetin, 143
synthesis, 117
in wine production, 142
flavan-3-ols, 95, 109, 114, 117, 118–119, 122, 141, 143
foliar analysis, 26
foliar fertilization, 10–14

G

gallic acid, 114, 128, 143
grapevine, 1–90
biogenic element, 15, 50
effect of soil pH on availability of nutrients, 18
endogenous factors, 21–24
exogenous factors, 24–26
fertilizer applications, 7–10
foliar fertilization, 10–13, 22
historical development, 2–4
microelements content, 10–15
macroelement content, 10–13, 15
nitrogen entry mechanisms, 18
nutrient requirements of, 7–8, 15–16, 19
physiological role of macroelements, 28–61
physiological role of microelements, 61–90
production, 4–5, 32
seasonal growth phases of, 9
significance of the vineyard growing, 4–14
utilization, role and significance of nutrient elements, 14–28
visual deficiencies, 12
see also Sulphur (S)
grapevine cultivation, 2–3, 16
grapevine leaves, 17, 93, 95, 114

grapevine nutrition, x, xi, 1, 8, 14, 26, 36, 43, 73, 84
assimilation of nutrients, 16–20
balance and antagonism, 26–28
copper deficiency, 73
factors influencing ion accumulation, 21–26
molybdenum deficiency, 84
movement of nutrients, 20–21
role of potassium, 43
utilization, role and significance, 14–16

I

ion synergism, 28
iron (Fe)
absorption, 65
deficiency, 52, 65–67, 71, 81–82, 85. *see also* iron deficiency
excess of, 68
role in metabolic process, 65–66
symptoms of deficiency and excess, 67

K

Krebs cycle, 39, 68

L

lipids, 5, 92, 94, 105–106, *106*, **108**

M

maceration process, 114, 120
macroelements, 13, 14, 15, 28, 28, 56, 91
macroelements, physiological role, 28–61
calcium (Ca), 47–50
carbon (C), 29
hydrogen (H), 29
magnesium (Mg), 50–54
nitrogen (N), 29–38
oxygen (O), 29
phosphorus (P_2O_5), 38–41
potassium (K_2O), 42–47
sulphur (S), 54–61
symptoms of deficiency and excess of macroelements, 56–61
magnesium (Mg), 50–54
absorption and transport, 50–51
biogenic element, 50
deficiency, 51–54, 60–61
excess of, 51, 54, 82
in grapevine, 50–54
greater mobility, 50
harmful effects, 54
mechanical composition, 51
photosynthetic pigments, 67
physiological functions, 50
symptoms of deficiency and excess, 52
see also Krebs cycle; molybdenum; phytosterols
malic acid, 5, 35, 98, 100
malvin, 122
manganese (Mn), 68–71
absorption, 68
chlorosis, 52, 71
biocatalyst, 69
deficiency, 52, 68, 70–71, 81, 86
excess of, 71, 84
in photosynthesis, 69
role of biocatalysts, 69
uptake by grapevines, 69
see also chlorophyll; molybdenum
metabolites, 5, 30, 92, 95, 147
microelements, ix, x, xi, 1, 10, 11, 12, 13, 14, 15, 20, 26, 41, 42, 49, 61, 62, 63, 64, 65, 85, 91, 103
microelements/micronutrient, physiological role, 61–90
biocatalysts role, 62
boron (B), 74–78
cobalt (Co), 82
copper (Cu), 71–74
deficiencies, 64
effect on pollen germination and fertilization, 63
iron (Fe), 65–68
manganese (Mn), 68
molybdenum (Mo), 82–84
in oxidation-reduction processes, 61–62
for plant growth, metabolism and symptoms associated with deficiencies, 62
relationship between soil pH and, 63
role in biochemical processes, 61
zinc (Zn), 78–82

microorganisms15, 25
minerals, 102–104
  cations and anions, 102
  vitamins and, 104
molybdenum, 15, 27, 41
  absorption, 82–83
  chlorosis, 84
  deficiency, 84
  excess of, 84
  increase in osmotic pressure, 84
  oxidation-reduction processes, 82
  physiological roles, 84
  symptoms of deficiency and excess, 83
Muscat grape, 136
myricetin, 117, 135, 143, 146

N

nitrogen (N), 29–38, 104–105
  agronomic practices, 104
  changes in N content of plant parts, 32
  compounds, 5, 34, 72, 104–105
  deficiency, 35–36, 55, 56, **62**
  excess of, 36–38, 57
  impacts of environmental conditions, 30
  redistribution, 33–34
  symptoms of deficiency and excess, 38
  uptake and assimilation, 33
  vegetative versus reproductive development rates, 31
  see also chlorosis; photosynthesis
NPK fertilizers, 64
nutrient deficiency, 7, 8, 64, 73
nutritional value of grapes, juice and wine, 108

O

oxidation-reduction processes, 13, 61, 68, 82

P

PAL activity, 101–102
pesticides, 4, 8, 73
phenolic acids, 95, 109, 115, 144
  antioxidant activity, 115
  grapevine leaves, 94, 95, 114
phenolic compounds104, 109–116, 123, 130, 138, 141. *see also* phenolic acids
  anti-inflammatory properties, 144
  and biological potential of wine, 138–148
  chemical structure, 111, 114
  classes of, 110, 112
  distribution, 114
  metabolism of, 114
  of red grape, 111
phenols, 109–110, 113, 117
phospholipids, 106
phosphorus (P_2O_5), 38–41
  accumulation of, 23, 25
  antagonism, 27–28
  deficiency, 40, 41–42, 57–58
  energy transfer, 39, 40
  excess of, 42
  grapevine, 38–39
  physiological and metabolic processes, 39–40
  root formation, 23
  symptoms of deficiency and excess, 41
photosynthesis, 31, 32, 39
  nitrogen deficiency and, 35
  phosphate, 39
  potassium, 43, 44
  role of calcium, 48
  role of manganese, 69–70, 71
    physiological role, x, 1, 28, 39, 43, 47, 48, 49, 61, 84
  macroelements, 28–61
  microelements, 61–90
phytoalexins, 95, 115
polyphenolic compounds, in grapes, 5, 92, 107, 108–129, 145, 147
  anthocyanins, 119–124
  flavan-3-ols, 118–119
  flavonoids, 116–117
  flavonols, 117–118
  phenolic, 109–116
  proanthocyanidins, 119
  tannins, 124–129
polyphenols, 6, 108–109, 119
polyvalency, 71, 82
potassium (K_2O), 42–47
  cluster growth, 44
  deficiency, 45–47, 58–59
  enzyme activation, 43
  excess of, 47

flowering, 44
functions in grapevine, 43
in grapevine nutrition, 42–47
physiological maturity of the grapes, 44
regulating the permeability of living membranes, 43–44
symptoms of deficiency and excess, 46
uptake of, 42
proanthocyanidins, 113, 116, 119, 123, 125, 139–140, 144

Q

quercetin, 117, 135, 142–143, 146

R

red wines, 139
alcoholic beverages, 144
anthocyanins and tannins, 141
catechin, 141
chemical components, 140
constituents, 145
intensive macerations, 130
malvin, 122
myricetin, 117, 135, 143, 146
phenolic content, 109, 142
production, 139
red-coloured polymers, 130
resveratrols, 6, 109, 111, 115–116, 147

S

stilbenes, 32, 95, 109, 115–116, 144
antimicrobial activity, 95, 115
bacterial and fungal infections, 115
derivatives, 115
phenolic compounds, 116
phytoalexins, 115
sulphur (S), 54–61
deficiency, 55
excess of, 56
role in physiological processes of grapevine, 54
symptoms of deficiency and excess, 55
see also grapevine; sulfur-based sprays

T

tannins, 124–129
and anthocyanins, 124
condensed, 124, 125, 127
formation of, 127
from grape skins and seeds, 126, 128
water-soluble, 124
in wine, 125
tartaric acid, 98

V

vineyards
factors influencing growth and yield, 1
K-deficient, 47
long-term viability of, 9
management practices, 1, 2
N fertilizer management, 7, 23
nutrient deficiency, 8
nutritional requirements of, 8, 13
role of fertilization, 8
significance, 4–14
viniferins, 116
vitamins, 104, 107–108
viticulture, 3–4
cultivation of vines, 1
economic characteristics, 4
fruit quality, 7
phosphorus, 39
production, 4, 7–8
tropical and subtropical regions, 23
and winemaking, 1

W

waxes, 94, 105–106, 108
white wine
antioxidants in, 141
flavonols, 141
kaempferol and quercetin, 143
phenolic composition, 142
production, 130, 134
wine
alcoholic beverage, 139
biological potential of, 138–148
cardiovascular disease, treatment of, 139, 144
consumption of, 139

flavonols in, 142
neurodegenerative diseases, treatment of, 146–147
phenolic compounds, 138–148
red *see* red wine
winemaking, 1
mineral matter, 103
nitrogen in, 34
pectolytic enzymes, 101
proanthocyanidins, 119, 120

X

xylem, 12, 48, 50, 74–77

Z

zinc (Zn), 78–82
absorption, 78, 80
affecting the oxidation-reduction, 13, 61, 68, 82
biosynthesis of DNA and RNA, 79
deficiency, 80–81, 89
excess of, 28, 82
in grapevine, 80
symptoms of deficiency and excess, 80
uptake by grapevines, 79
see also grapevine; molybdenum; phytosterols